Research Series on the Chinese Dream and China's Development Path

Series Editors

Yang Li, Chinese Academy of Social Sciences, Beijing, China

Peilin Li, Chinese Academy of Social Sciences, Beijing, China

Drawing on a large body of empirical studies done over the last two decades, this Series provides its readers with in-depth analyses of the past and present and forecasts for the future course of China's development. It contains the latest research results made by members of the Chinese Academy of Social Sciences. This series is an invaluable companion to every researcher who is trying to gain a deeper understanding of the development model, path and experience unique to China. Thanks to the adoption of Socialism with Chinese characteristics, and the implementation of comprehensive reform and opening-up, China has made tremendous achievements in areas such as political reform, economic development, and social construction, and is making great strides towards the realization of the Chinese dream of national rejuvenation. In addition to presenting a detailed account of many of these achievements, the authors also discuss what lessons other countries can learn from China's experience.

Project Director

Shouguang Xie, President, Social Sciences Academic Press

Academic Advisors

Fang Cai, Peiyong Gao, Lin Li, Qiang Li, Huaide Ma, Jiahua Pan, Changhong Pei, Ye Qi, Lei Wang, Ming Wang, Yuyan Zhang, Yongnian Zheng, Hong Zhou

More information about this series at http://www.springer.com/series/13571

Binbin Wang

China's Transition on Climate Change Communication and Governance

From Zero to Hero

Binbin Wang
School of International Studies
Peking University
Beijing, China

This research is funded by the Chinese Fund for the Humanities and Social Sciences (Project No.: 18WJY010)

ISSN 2363-6866 ISSN 2363-6874 (electronic)
Research Series on the Chinese Dream and China's Development Path
ISBN 978-981-15-8831-0 ISBN 978-981-15-8832-7 (eBook)
https://doi.org/10.1007/978-981-15-8832-7

Jointly published with Social Sciences Academic Press
The print edition is not for sale in Mainland China. Customers from Mainland China please order the print book from: Social Sciences Academic Press.

© Social Sciences Academic Press 2021
This work is subject to copyright. All rights are reserved by the Publishers, whether the whole or part of the material is concerned, specifically the rights of translation, reprinting, reuse of illustrations, recitation, broadcasting, reproduction on microfilms or in any other physical way, and transmission or information storage and retrieval, electronic adaptation, computer software, or by similar or dissimilar methodology now known or hereafter developed.
The use of general descriptive names, registered names, trademarks, service marks, etc. in this publication does not imply, even in the absence of a specific statement, that such names are exempt from the relevant protective laws and regulations and therefore free for general use.
The publishers, the authors, and the editors are safe to assume that the advice and information in this book are believed to be true and accurate at the date of publication. Neither the publishers nor the authors or the editors give a warranty, express or implied, with respect to the material contained herein or for any errors or omissions that may have been made. The publishers remain neutral with regard to jurisdictional claims in published maps and institutional affiliations.

This Springer imprint is published by the registered company Springer Nature Singapore Pte Ltd.
The registered company address is: 152 Beach Road, #21-01/04 Gateway East, Singapore 189721, Singapore

Series Preface

Since China's reform and opening began in 1978, the country has come a long way on the path of socialism with Chinese characteristics, under the leadership of the Communist Party of China. Over 30 years of reform, efforts and sustained spectacular economic growth have turned China into the world's second largest economy, and wrought many profound changes in the Chinese society. These historically significant developments have been garnering increasing attention from scholars, governments, and the general public alike around the world since the 1990s, when the newest wave of China studies began to gather steam. Some of the hottest topics have included the so-called "China miracle", "Chinese phenomenon", "Chinese experience", "Chinese path", and the "Chinese model". Homegrown researchers have soon followed suit. Already hugely productive, this vibrant field is putting out a large number of books each year, with Social Sciences Academic Press alone having published hundreds of titles on a wide range of subjects.

Because most of these books have been written and published in Chinese, however, readership has been limited outside China—even among many who study China—for whom English is still the lingua franca. This language barrier has been an impediment to efforts by academia, business communities, and policy-makers in other countries to form a thorough understanding of contemporary China, of what is distinct about China's past and present may mean not only for her future but also for the future of the world. The need to remove such an impediment is both real and urgent, and the *Research Series on the Chinese Dream and China's Development Path* is my answer to the call.

This series features some of the most notable achievements from the last 20 years by scholars in China in a variety of research topics related to reform and opening. They include both theoretical explorations and empirical studies, and cover economy, society, politics, law, culture, and ecology, the six areas in which reform and opening policies have had the deepest impact and farthest-reaching consequences for the country. Authors for the series have also tried to articulate their visions of the "Chinese Dream" and how the country can realize it in these fields and beyond.

All of the editors and authors for the *Research Series on the Chinese Dream and China's Development Path* are both longtime students of reform and opening and recognized authorities in their respective academic fields. Their credentials and expertise lend credibility to these books, each of which having been subject to a rigorous peer review process for inclusion in the series. As part of the Reform and Development Program under the State Administration of Press, Publication, Radio, Film, and Television of the People's Republic of China, the series is published by Springer, a Germany-based academic publisher of international repute, and distributed overseas. I am confident that it will help fill a lacuna in studies of China in the era of reform and opening.

Shouguang Xie

Foreword by Xiangwan Du

Climate change is a major global environmental issue and also a non-traditional security issue. Human understanding of climate change has been growing over the last two centuries or so. In general, the modern science of climate change has gone from theoretical investigations to empirical studies, and continued to mature and deepen as a field of scientific inquiry. Broad consensus is being reached over an increasing number of related issues. However, disagreements and skepticism still exist. Some even see it as a conspiracy and a trap. This also illustrates the necessity and significance of climate change communication. Much more needs to be done to communicate the science of climate change not only to the public, but also to decision-makers and even people of influence. Climate change communication has several goals. First, it should improve basic awareness of climate change as an issue facing all of humanity, increase understanding of the fact that behavioral changes would only happen after changes in beliefs and values, and facilitate realization of the need to build consensus on the basis of the science. Second, it should underscore the importance of adapting to climate change to humanity's shared interests, and the impact it could have on the transformation of development models around the world. Finally, climate change communication should promote joint adaptive actions, provide guidance on how to shift to a green, low-carbon and sustainable development path, and to transition from industrial civilization to ecological civilization.

The global governance in climate change requires the common actions of all humans, all countries and all nations, which should cooperate to share the results and experiences in addressing climate change and achieving sustainable development. In addition, the world should build a community of human destiny for sustainable development and climate change response together, which is a general trend.

China plays an important role in addressing climate change. Addressing climate change is the inherent demand of China for sustainable development. China is striving to improve its environmental quality just to address climate change, build a Beautiful China and realize the China Dream. In fact, China is a latecomer. However, as a responsible big country, China should make more contributions to the sustainable development of mankind and fulfil its solemn commitment to the world about addressing climate change. China is taking the initiative to address climate change. In conjunction with other countries, China will hold high the banner of "climate justice" and "win-win cooperation", address climate change properly, take the road of green and low-carbon development, contribute to and promote the establishment of an international new order, and make unremitting efforts for the sustainable development of the community of human destiny. No country is superior to other countries. All countries should work together to address climate change in the right way.

The UN's conferences in Copenhagen and Paris successively witnessed the global governance process of historical significance in climate change. Through the joint efforts of various countries, the Paris Agreement was finally concluded, which is a milestone in the global governance of climate change, and also a major achievement reflecting the rational concept of "co-shared community of human destiny". Although the world is full of differences, contradictions and even conflicts, people living on the same Earth have realistic and potential common interests and need to establish a new international order and a mechanism for co-construction, co-governance, and sharing.

Dr. Binbin Wang, the author of this book, has witnessed this process. As an enthusiastic young scholar, she has engaged in innovative research and practice in climate change communication and governance for many years, having done a lot of useful work. After the interview with me on the low-carbon transformation in 2013, she presided over a survey on the low-carbon cognition of Chinese urban residents. In mid-2016, I attended the Seminar on Global Climate Governance: New Situations, New Challenges and New Ideas held in Peking University. Dr. Wang told me that she planned to make a survey on China's changes in climate change response from the perspective of cooperative governance. Thinking it very valuable, I encouraged her to do that. In October 2017, Dr. Binbin Wang released the survey report on Climate Change in the Chinese Mind in Beijing. I was invited to write a preface for her survey report. Only half a year later, she published this book, which is another gift for the climate change researchers and practitioners after 2017, and enables you to review, rethink and look forward.

I hope more readers can benefit from this book.

March 2018

Xiangwan Du
Honorary Director of the National Expert
Committee on Climate Change of China
Academician of Chinese Academy of Engineering
Beijing, China

Foreword by Baowei Zheng

At present, the irreversible and serious damage caused by climate change to the natural environment and human society has posed the greatest threat to human survival and development. To change this situation, global co-governance is needed, on which most countries in the world have reached a consensus. The Paris Agreement adopted at the 21st UN Climate Change Conference in Paris (COP21) in 2015 is an official declaration of the international community to jointly address climate change.

The international community shoulders the same mission to address climate change. As a responsible big country, China should play more roles and make greater contributions in this regard. As President Xi Jinping emphasizes, China will play the role of "participant, contributor and driver" well.

Addressing climate change is related to the interests of the country, the nation, the society, and social members. To jointly address climate change is an inevitable choice of the whole society. To effectively address climate change, especially to form a social consensus, all people must be mobilized to get involved. In recent years, we have been advocating the construction of an action framework of five behavioral subjects (including the government, media, NGOs, enterprises, and the public) for climate change response and climate change communication, to make addressing climate change truly become social consensus and universal action respectively.

The so-called "climate change communication" is a type of social communication activity aimed at helping the public better understand climate change-related information and scientific knowledge, and adopt environment-friendly attitudes and behaviors. It is meant to be an integral part of a larger solution to this global problem. Simply speaking, climate change communication is a social communication activity related to climate change information and knowledge and aiming to solve the problems of climate change.

Climate change communication research mainly focuses on the following issues: definition of climate change communication; relationship between climate change communication and climate change; behavioral subject and role of climate change communication; national strategy and action strategy of climate change communication; advocacy and public participation; international exchange and cooperation; skills and techniques.

Climate change communication research requires the systematic exploration of climate change communications, promotion of related practice, studying the significance of energy conservation, emissions reduction and environmental protection to the economic and social development of the country and human being, and the theoretical consultation and academic support for the government, media, enterprises and NGOs in climate change communication, etc.

In April 2010, with the support of all parties, China Center for Climate Change Communication (China4C), the first local think tank in the field, was established. Through a few years of hard work, China4C has continuously enriched the research content, expanded research methods and results, and won the attention and recognition of academic circles and related departments at home and abroad, arousing extensive coverage by a large number of media.

Since its establishment in 2010, China4C has completed a series of research and advocacy work at home and abroad.

Participation in the UN Climate Change Conference and Hosting the Climate Change Communication Conference

Since 2010, China4C has successively participated in the UN Climate Change Conferences in Cancun, Durban, Doha, Warsaw, Lima, Paris, Marrakech, Bonn and Katowice, and "Rio+20" UN Conference on Sustainable Development in 2012, having participated in the global climate governance in key stages.

Since the establishment of "China Pavilion" at the UN Climate Change Conference, China4C has held a number of side events themed by "climate change communication and public participation", and built the platform for domestic and foreign scholars, officials, media organizations, and civil society to exchange experience, display research results and express stances and views. This platform with certain influence at home and abroad has been recognized and affirmed by many friends in the field.

Organization of International Academic Conferences on Climate Change Communication

In recent years, we have held several conferences on climate change communication. For example, in 2010, we held a Seminar on Climate, Communication, Interaction and Win-Win—The Role and Influence of Government, Media and NGOs in the Post-Copenhagen Era. This is the first seminar organized by China4C, at which the invited domestic and foreign scholars, media and NGOs, as well as representatives of government agencies summarized the experience and lessons of governments, media and NGOs in climate change communication during the COP15 in

Copenhagen in 2009, and put forward suggestions and opinions from the perspective of academic research. We proposed the concept of "climate change communication", starting the theoretical research on climate change communication in China.

In 2013, we hosted the 2013 International Conference on Climate Change Communication in cooperation with the Yale Center for Climate Change Communication (YPCCC), which is the first largest and most influential international conference on climate change communication so far. A lot of consensuses were reached on the ways of climate change response and communication, which, to some extent, advances the theoretical research and advocacy of climate change communication.

In 2016, we held the Seminar on Green Development and Climate Change Communication, exploring how to promote climate change communication and green development.

Conducting Surveys on Public Awareness of Climate Change and Climate Change Communication in China

In order to accurately understand the public perceptions of climate change, so as to make the more targeted climate change information communication, knowledge popularization and social mobilization, China4C, in 2012 during the 12th Five-Year Plan period, organized the first survey on China's public awareness of climate change and climate change communication. The survey results were bought in by the White Paper: China's Policies and Actions on Climate Change, and quoted and affirmed by Christiana Figures, the then Executive Secretary General of the Secretariat of UN Framework Convention on Climate Change.

In 2013, China4C launched the Survey on Low Carbon Awareness and Behavior of Chinese Urban Public in line with China's first "Low Carbon Day" event, and released the survey results at the Diaoyutai State Guesthouse.

In 2017 during the 13th Five-Year Plan period, China4C conducted the second survey on China's public perception of climate change with the same method and number of sampling to conduct the comparative research with the one five years ago. We held the release conference in Beijing and COP23 in Bonn, Germany, respectively. Patricia Espinosa, the incumbent Executive Secretary General of the Secretariat of UN Framework Convention on Climate Change (UNFCCC) spoke highly of our report and sent a spokesperson Nick Nuttall to participate in our press conference in Bonn. Survey results show that 96.8% of the respondents support the Chinese government's international cooperation in addressing climate change. Mr. Jorge Chediek, Special Envoy of the UN Secretary-General for South-South Cooperation, regarded the report as "a piece of good news for the world".

Engaging in Climate Change Advocacy

In 2012, China4C shifted the focus of academic research to social mobilization and promotion, and launched the 4-dimension (improve perceptions of climate change in the communities, campuses, enterprises and rural areas) climate change communication campaign around the country, advocating the establishment of an action framework of five behavioral subjects (including the government, media, NGOs, business, and the public) to promote the active participation of the public.

China4C also worked with other universities and colleges to promote the research and practice on climate change communication. At present, Communication University of China, Xinxiang University in Henan, Qingdao University, South-Central University for Nationalities, Guangxi University and other universities have established or are setting up climate change communication center. We hope more and more people can participate in theoretical research and advocacy of climate change communication so as to make it have larger influence.

Publications

We were also involved in academic research. In addition to related theme papers published in core journals, *The Theory and Practice of Climate Change Communication*, China's first monograph on climate change communication, was published at the end of 2011. At the COP17 in Durban that year, the version of this book in both Chinese and English won high praises from experts in the field.

In 2014, commissioned by the Euro-China Forum, China4C presided over the drafting of the *Consensus of China and Europe on Addressing Climate Change*, which conveys the voice of Chinese and European people for the COP21 to be held in Paris in 2015.

In 2015, China4C translated and published the *Climate Change CommunicationPsychology* compiled by Columbia University.

In 2017, the book *Green Development and Climate Change Communication* was published to introduce the latest achievements in the climate change communication field.

In recent years, China4C has been working hard to make the research team stronger through talent cultivation and team building, so as to improve the research level, expand the academic influence, and truly form an atmosphere of climate change communication in China.

Dr. Binbin Wang is a co-founder of China4C and the pioneer in the field. Her new book *From Zero to Hero: China's Transition on Climate Change Communication and Governance* will soon be published. She is my first doctoral student to write a dissertation in climate change communication, and also the first one to earn a Ph.D. in this field in China. Committed and driven, she conducted interdisciplinary research that matched the broad range of frontline practices she was engaged in. This book is just an epitome of her years of efforts. In this book, she defines the logical relationship between climate change communication and climate governance, clarifies the strategic significance of climate change communication, and summarizes China's path of participation in climate governance from the perspective of multi-stakeholder cooperation.

At the Symposium on Philosophy and the Social Sciences in 2016, Chinese President Xi Jinping stressed the importance of furthering research in these fields and called on the academic community to help the world better understand China's progress, commitment to openness and capability to contribute to human civilization.

Dr. Binbin Wang's new book presents her insights about China's experience in global climate governance. It will do more than informing her readers, it will also inspire them to do their part.

March 2018

Baowei Zheng
Founder and Director of China Center
for Climate Change Communication
Renmin University of China
Beijing, China

Preface

At the end of August 2016, as the China Climate Lead of Oxfam, an international developmental and humanitarian NGO, I invited Minister Zhenhua Xie, China Special Representative on Climate Change Affairs to visit Oxfam's field project in Vietnam. This visit aims to understand the real demands of people in developing countries through Oxfam's network, so as to better design the usage of China's South-South Climate Cooperation Fund with a total amount of USD $31 billion.

Vietnam is one of the five countries in the world that are most affected by climate change. Climate change and rising sea levels have given rise to flooding and seawater intrusion, thus causing heavy losses to Vietnam's economic and social development and local people's life. The impacts of climate change are evident in the densely populated Red River Delta and the Mekong River Delta, which are mainly reflected in extreme flood disaster, anti-season temperature change, sea level rising, seawater intrusion, and tropical storm, etc. The extreme weather events caused by climate changes have seriously affected the livelihoods of local villagers, resulting in the shrinking of agricultural land and clean water sources, the erosion of coastal areas, and the degradation of ecosystem.

After three hours of driving, we arrived at the Giao Xuan commune in Nam Dinh seriously affected by climate change. The 3 km mangrove forest under the jurisdiction of Giao Xuan commune has been reduced year by year due to seawater intrusion and man-made damage. The local government bans some environmental disruption behavior of local villagers in the process of fishing and farming, which worsens the relationship between them. In order to solve this problem, the local government, communities and NGOs set up a joint management team and signed protection agreements with local villagers on the principle of voluntariness, to jointly manage natural resources and seek development.

Due to seawater intrusion, original paddy fields were submerged somewhat, which seriously affected the lives of local villagers. After understanding the actual needs of local villagers, the project provided alternative choices from growing to cultivating, and co-designed the development of eco-tourism and establishment of organic living cafes, which has improved the livelihood of local villagers.

After the visit, Minister Xie communicated with the villagers sitting around.

"Your experience shows that mitigation and adaptation must be combined well with poverty alleviation, economic development and protection of local ecological environment, to improve the people's livelihood and achieve the sustainable development."

Minister. Xie also stressed: "This project is a very good example of joint engagement of government, international organizations, local NGOs, private sectors and villagers."

During the dinner, Minister Xie asked me: "Why do you want to promote the multi-stakeholder cooperation?"

I answered unconsciously: "This is my job."

He said looking at me: "Addressing climate change is not just a job, but a cause."

After this visit, I buried myself inself-reflection. I have gained much growth, trust and support in the past eight years. Since it is a cause, what else can I do?

I got involved in climate change work for the first time in 2009.

On December 7, 2009, the COP15 was held in Copenhagen, Denmark, which is the largest international negotiation in history and a milestone event in the history of global governance. I witnessed the full process of this historical event, the efforts of Chinese government to reach an effective agreement, the efforts and helplessness of Chinese media, and also the global network and influence of various intergovernmental organizations and NGOs. However, ten years ago in Copenhagen, there was basically no dialogue between them.

The first-line observation of Copenhagen Negotiations made me begin to think over the roles and strategies of three major stakeholders, namely, government, media and NGOs, hoping to find a path of win-win cooperation in addressing climate change. In the process of follow-up negotiations, I conducted the tracking analysis successively and observed the transition of China from the two-level game perspective.

Domestically, with the frontier knowledge I learnt from the international peers in different side events of COPs, I worked with my team and partners to carry out the comprehensive pilot work about low carbon adaptation and poverty alleviation in a small poor village in Shaanxi Province of China. Shaanxi is in the middle of the country, and frequent and unregular extreme weathers aggravate the negative climate impacts on local small farmers' livelihood. It aims to explore a feasible approach that combines climate change adaptation, low carbon development, and poverty alleviation for sustainable development in rural communities, through interaction and cooperation with the government at all levels, research institutes, non-governmental organizations (NGOs), the private sector and the media. A gender-sensitive participatory approach has also been incorporated into the programme. After nearly two years of implementation, the target group has gradually come to better understand climate change and realized that low carbon development is not an economic burden, but a concept and lifestyle that enhances agricultural production and rural life. They also learnt that poverty reduction and low carbon adaptation are complementary. By introducing low carbon technology for both agricultural production and community life, the programme could greatly reduce greenhouse gas emissions. More specifically, it would be reduced by a total of 2,249 tCO2, or 14 tCO2e a year per household. Through the programme, annual income for locals increased from RMB 2,100 to 3,500 yuan within less than two years, between 2013 and 2015. As the programme continues, it is expected that annual incomes will continue to gradually increase. Just before the COP21 at the end of 2015, this pilot was awarded "Pioneer of Change", one of the best practices of low-carbon sustainable development in developing countries. I took the lessons learnt from the process back to international platform to share and exchange, and contribute to the improvements of the relative solutions. In this way, international peers further understand China's actual situation and experience, increasingly realizing that international experience is not a universal panacea.

In the process of two-level communication and personal participation, I realized the interdisciplinary and cross-border characteristics of climate change issues, the two-level game and interaction at both international and domestic levels, and the importance of multi-participation in governance, striving to promote multi-dimensional cooperation. In this process, I saw that Chinese government, media and NGOs became allies of jointly addressing climate change.

The COP21 in Paris at the end of 2015 is another milestone event for global climate governance, at which the historic Paris Agreement was concluded. Only through six years of efforts, China became a leader from a follower.

After returning from Paris, I, invited as a spokesperson, participated in the press conference with Chinese and international media hosted by China's State Council Information Office. Minister Zhenhua Xie, the then China's Special Representative on Climate Change Affairs, Mr. Wei Su, the then director of the Climate Change Department of the National Development and Reform Commission, Mr. Ji Zou, the then Deputy Director of the National Center for Climate Change Strategy and International Cooperation, Ms. Xiaoming Zhu, the then Mayor of Zhenjiang City, and Mr. Wenbiao Wang, Chairman of Yili Group attended this event as spokesperson together with me.

From the perspective of governance, this is the first time that China shows its inclusive and open climate governance pattern of multi-participation to the world through the national news release platform since the participation in global climate governance.

In history, this should be the first national press conference witnessed by multiple stakeholders, especially representatives of NGOs.

In 2017, President XI Jinping, in the Report at the 19th Party Congress, proposed to strengthen and innovate social governance, improve the community governance system, promote the shift of social governance focus to the grassroots level, give play to the role of social organization, and build a new pattern of co-construction, co-governance, and sharing.

I realized the experimental field role of climate governance in social co-governance. I have witnessed the rapid change of China in the past few years.

Minister Xie's sincere words in Vietnam made me feel necessary to share my experience and research results with more people, so as to make more people interested in and confident about this cause of climate change.

On the basis of my doctoral thesis, I construct a "two-level, multi-dimensional" research space, making the case analysis, and revealing China's climate governance transition process. I think that China's participation in the global climate governance is innovative and based on successful explorations at both the international and domestic levels, and is the best example of social co-governance.

My research in Chinese version has been subsidized by the Natural Science Foundation Emergency Management Project entitled "Impact of the U.S. Withdrawal from the Paris Agreement on Global Climate Governance and China's Response Strategy". This project requires an assessment of the impact of U.S. withdrawal from the Paris Agreement on all parties. There are two main findings as follows.

First, even the U.S. has withdrawn from the Paris Agreement, but the American people are still in!

Second, although the top-down governance mode is undergoing the transformation, bottom-up innovation momentum has burst out.

As long as we continue to adhere to the concept of social co-governance and encourage multiple participation, the torch of global climate governance will not be extinguished.

This book records the glory and legend of China in participation in global governance in the past 40 years of reform and opening up. I am fortunate to win the trust and support from many colleagues. In the past eight years, we, despite different units and different backgrounds, joined hands in climate governance. Whenever it occurs to me, I feel warm and grateful.

Here, I would like to express my special thanks to Minister Zhenhua Xie. With the strong personality charm and spiritual strength, he brought together stakeholders in the process of climate governance. I believe that everyone will remember those unforgettable days full of ups and downs. Fortunately, I was encouraged and reminded by Minister Xie several times, deeply feeling his broad vision and strategic strength, which injected power for my growth.

I hereby extend sincere gratitude to Academician Xiangwan Du and Professor Baowei Zheng for writing the preface for this book. Both of them are over 70 years old, but they still work tirelessly, actively participating the UN climate change process, spreading the scientific knowledge of addressing climate change, and calling for addressing climate change.

Academician Du always stresses the necessity to spread the climate change science better and more widely, analyzing the scientific cognition of climate change, the importance of global climate governance, China's role in global climate governance, and new order of global climate governance. I believe that all those who have read this preface carefully will have a new understanding of them.

When returning from Copenhagen in 2009, I exchanged views with Prof. Baowei Zheng, who have something in common with me on the negotiation effects. Since then, we have jointly started the study and practice of climate communication and cofounded the China Center for Climate Change Communication (China4C) in 2010. In the preface, Prof. Zheng systematically reviewed the five major aspects of work of the China4C at home and abroad. Climate communication is a strategic task of climate governance. Although the public has a high awareness of climate change, they still lack the willingness to take practical actions. We need to explore the solutions to the bottleneck of climate communication. After reading Prof. Zheng's preface and this book, if you have any idea about using innovative methods to narrow the gap between cognition and action, please contact us.

I also need to express my gratitude to School of International Studies, Peking University, for the research environment and rich spiritual nourishment. In particular Prof. Haibin Zhang, who has more than 20 years of experience in global environment and climate governance in China. He has always maintained academic enthusiasm and humanistic care in this field and encouraged me to go ahead by following my heart.

The publication of the Chinese version of this book was supported by the editor Jingyi Wang and the Social Sciences Academic Press. With their recommendation, the English version can be published by the Springer. Ms. Yingying Zhang is the editor of this book in English and encouraged me a lot in the process. I would like to express my gratitude to them.

In the past eight years, I have gained supports from many colleagues in different organizations but caring about the same cause of climate. I hope to present my book to them to come up with new inspiration.

Since 2009, China have been practicing the two-level climate co-governance in an innovative way. I hope this book can enable more readers to see China's path and the hope, faith and possibility of the common cause. We only have one planet and we should come together, without prejudgment, discrimination, and isolation, to address climate change and protect our home for a sustainable future.

Beijing, China   Binbin Wang

Recommendations

Recommendation 1

From 2009 to date, China has embarked on a path of win-win cooperation in global climate governance. This book is a useful reference for all parties to understand this process.

XIE Zhenhua
China's Special Representative on Climate Change Affairs
03.20.2018

Recommendation 2

One of the main features of the book is an in-depth analysis of the role and strategic shift of climate communication and governance between 2009 and 2015 by China's government, media and NGOs, and it makes an innovative exploration of the Chinese path of understanding global climate governance.

Haibin Zhang
Vice Dean and Professor
School of International Studies at Peking University

Recommendation 3

This book chronicles the China's passive-to-active approach to global governance of climate change, and its future posture leading the way in transforming and promoting multi-interactive action in the two-level game. The climate-field NGOs in China are making a leap forward. The most valuable thing is that the author, with the enthusiasm of the actors and the wisdom of the researchers, sums up the experience from practice, links theory with practice, and provides ideological guidance to her peers.

Lo Sze Ping
China Representative of WWF

Contents

1 **Introduction** 1
   1 Research Background 1
   2 Theoretical Foundation 2
      2.1 Governance and Global Climate Governance 2
      2.2 Two-Level Game Theory 8
   3 Research Methods 10
      3.1 Two-level Analytic Hierarchy Process 10
      3.2 Analysis of Stakeholders 10
      3.3 Tracking Analysis 12
   4 Research Framework and Value 13
      4.1 Research Framework 13
      4.2 Research Value 14

2 **Climate Change and Climate Change Communication** 17
   1 Summary of Climate Change Research 17
      1.1 Certainty of Climate Change 17
      1.2 Uncertainty of Climate Change 19
      1.3 Precautionary Principle 20
   2 Summary of Climate Change Communication
     Research 23
      2.1 Definition of Climate Change Communication 23
      2.2 Relationship Between Climate Change Communication
         and Six Major Application Communication Modes 24
      2.3 International Research on Climate Change
         Communication 31
      2.4 China's Researches on Climate Change Communication ... 36

3 **Two-Level Games of China** 43
   1 Motivation 43
      1.1 Domestic Roots of International Problems 44

		1.2	International Roots of Domestic Problems	46
	2	Objects		47
		2.1	International Game Objects	47
		2.2	Domestic Game Objects	48
		2.3	Adaptative Revision of Win-Set at the Domestic Level	49

4 Analysis of Stakeholders (2009–2015) 51
  1 Definition and Classification of Stakeholders 51
      1.1 Definition 51
      1.2 Classification Method 52
  2 International Stakeholders 53
      2.1 Legitimacy 53
      2.2 Power 54
      2.3 Urgency 55
      2.4 Correlation 56
  3 Domestic Stakeholders 57
      3.1 Legitimacy 57
      3.2 Power 58
      3.3 Urgency 59
      3.4 Correlation 60
  4 Three Major Stakeholders: Governments, Media and NGOs 60

5 Empirical Study: Two-Level Tracking Analysis of Three Major Stakeholders (2009–2015) 63
  1 Role Changes of Chinese Government 64
      1.1 Role of the Chinese Government in COP15 64
      1.2 Tracking Analysis: Changes in Government Strategies at International Level 70
      1.3 Tracking Analysis: Changes in Government Strategies at Domestic Level 74
  2 Role Changes of Chinese Media 77
      2.1 Role of the Chinese Media in COP15 78
      2.2 Tracking Analysis: Changes in Media Strategy at International Level 80
      2.3 Tracking Analysis: Changes in Media Strategy at Domestic Level 82
  3 Role Changes of NGOs 86
      3.1 The Role of NGOs in COP15 86
      3.2 Tracking Analysis: Strategy Shift of NGOs at the International Level 90
      3.3 Tracking Analysis: Strategy Shift of NGOs at the Domestic Level 92
  4 Main Conclusions 95

6	**China's Climate Change Communication and Governance in the Post-Paris Era**		99
	1 China's Challenges		99
		1.1 Continuous Enhancement of Right of Speaking	99
		1.2 International Expectations and China's Action	100
		1.3 The Largest Win-Set not yet Formed	101
	2 China's Strategy		102
		2.1 Telling Stories of "Real China" from Domestic to International	102
		2.2 "Pressure Transmission" from International to Domestic	103
		2.3 Strategy of Cross-Level "Collaborative Governance"	105
	3 Case Study		108
	4 Case Study		110
7	**New Stage of "Dual Transition" (2015–2018)**		113
	1 "Dual Transition" Stage of Global Climate Governance		114
		1.1 Transition of the Leadership	114
		1.2 Transition of Emissions Reduction Mode	118
	2 Opportunities and Challenges for China at the New Stage of "Dual Transition"		120
		2.1 Opportunities	121
		2.2 Challenges	124
	3 Strategic Choice		125
		3.1 Strategic Choice of China's Climate Governance	125
		3.2 Strategic Response for China's Climate Change Communication	128
8	**Conclusions and Outlooks**		131
	1 Major Conclusions		131
		1.1 Theoretical Levels: The "Dual-Layer, Multi-dimensional" Research Framework	131
		1.2 Practice Level: China's Route Choice	134
	2 Outlook		135
		2.1 From Top to Bottom: Way of Reform of Regime Complex	135
		2.2 From Bottom to Top: New Drive for Local Climate Change Response	136

Appendix 1: Climate Change in the Chinese Mind 2012 145

Appendix 2: Climate Change in the Chinese Mind 2017 169

References 203

About the Author

Dr. Binbin Wang has devoted her career to leadership on achieving sustainable development goals, bringing decades of expertise and experience in the synergistic effect between climate change and other SDGs, i.e. poverty alleviation, biodiversity, economic development and social justice, etc. Dr. Wang is currently Executive Secretary of Global Climate Change and Green Development Fund, an initiative from Mr. Xie Zhenhua, China's Special Representative on Climate Change Affairs, to mobilize the social capital and investment to support the global green transformation trend and seating in the core leadership team of Institute of Climate Change and Sustainable Development at Tsinghua University. She is also a Research Fellow of School of International Studies at Peking University. Prior to her current role, she spent eight years with Oxfam, one of the most influential international humanitarian and developmental organizations, and led the team to advocate policy makers in China to consider the benefits of the vulnerable people through setting up a bridge between climate change and poverty alleviation. Dr. Wang hosted a pilot of the Low-Carbon Adaptation and Poverty Alleviation Plan (LAPA) in a small village in the Shaanxi Province from 2012 to 2015, this pilot was then awarded the 'Pioneer of Change', one of the 21 Best Cases on Low Carbon Sustainable Development in Developing Countries.

As the first Ph.D. on climate change communication in China, Dr. Wang co-founded the local think tank based in Beijing a decade ago, the China Center for Climate Change Communication, which hosted national public perception surveys to evaluate the public opinions on climate. The key findings were recorded on China's Annual National White Paper on Climate Change, which then was used by the UNFCCC. Leveraging public opinions, Dr. Wang successfully advocated the Chinese government to open dialogue with civil society and engage the voice of people in the process of climate legislation. Because of Dr. Wang's remarkable contribution, she, as the first spokesperson on behalf of the civil society in the history of China, was invited to participate the press conference hosted by the China State Council right after COP21 to share her professional opinions on the importance of multi-stakeholder cooperation. With her experiences in the field, Dr. Wang was once invited as the Special Advisor for UN's Office on South-South Cooperation and is currently serving as a member of the Biodiversity Expert Group of China Council for International Cooperation on Environment and Development (CCICED), a high-level international advisory body hosted by China's State Council, to set up bridge between the CBD COP 15 in Kunming and UNFCCC COP 26 in Glasgow in 2021 to push the process of global governance ahead. In recognition of her remarkable contribution in the field, she was selected to represent female leadership in China for the "Homeward Bound", a transformational leadership initiative for women scientists and activists from around the world to explore the Antarctica in 2019.

Chapter 1
Introduction

1 Research Background

In the face of fluid international conditions, the Chinese government has expressed the intention to fully perform the responsibilities of China as a responsible big country, implement the "Belt and Road" Initiative, build a community of human destiny, and actively present Chinese solutions for global governance.

Chinese solutions to the global governance stem from Chinese experience and practice.

Global governance process is closely related to China's participation degree. In only six years from COP15 in Copenhagen in 2009 to the conclusion of the Paris Agreement in 2015, China completed the transition from the passive follower to active leader, successfully shaping an image in global climate governance. This is the first time that China moved to the center of the world stage at the fastest speed in the past 40 years since engagement in global climate governance.

What factors lead to the transformation? What experience can be summarized? What new situations and new possibilities are there after the U.S. announcement to withdraw from the Paris Agreement? What experience can be used by China for participation in other global governance issues, i.e. the implementation of the "Belt and Road" Initiative?

I'm eager to find the answers to the above questions.

In the past two years, China's academic circle and policy research circle have consciously started academic and realistic exploration of climate governance path, making related researches on international climate change response rules, relationship between China and global climate governance, climate strategies of key government behavior subjects, and special mechanisms for climate change negotiations, etc. Particularly, some researchers are government delegation members that have followed up the governance process for a long time and can make close-range researches on state actors. Such researches enable readers and peers to deeply understand China's relationship with global climate governance.

Apart from the above research, it is stressed that multiple actors should jointly address global common issues. From the COP15 in 2009 to COP21 in 2015, China experienced a "tough fight" in which state actors and many non-state actors got involved. At home and abroad, state actors and non-state actors interact more and more frequently, and their interaction is also worthy of study to sort out China's experience in engagement in global climate governance.

I have been engaging in the related work of climate change and sustainable development since 2008. In 2009, I, as the first Chinese representative of an international NGO, personally participated in the Copenhagen Conference, which benefited the Chinese government, media and NGOs a lot, inspiring me to start tracking analysis of role positioning and strategic change of governments, media and NGOs in climate governance at the international and domestic levels.

Based on the past ten years of participatory follow-up study and practice, as well as two-level game theory and global governance theory, I construct a "dual-layer, multi-dimensional" research space in this book.

After the empirical research on the strategy change and interaction of governments, media and NGOs from 2009 to 2015, I find the Chinese path of climate governance is double-leveled and multiple. I believe multi-level and multiple global governance is not impossible.

Global governance is a dynamic process and will not yield fruit only due to the conclusion of historic Paris Agreement at the end of 2015. While the world was still immersed in the joy of signing the Paris Agreement and preparing to implement it, newly elected U.S. president Trump, a climate change sceptic, announced the withdrawal from the Paris Agreement.

New changes bring new uncertainties to global climate governance. Based on the "dual-layer, multi-dimensional" research space and new observations in the last two years, I look forward to the future of global climate governance. The top-down mechanism complex reform is in progress, and more stakeholders including the general public and entrepreneurs (bottom-up) are doing pilot work about climate governance. I believe that a self-circulating "Convective Zone" of climate governance will be eventually formed.

2 Theoretical Foundation

2.1 *Governance and Global Climate Governance*

With the advent of globalization era, original nations, countries and private sectors cannot solve all problems through unilateral actions, thus governance models emerge. Compared to the traditional model of rule, governance refers to the activity for the common goal.

James N. Rosenau, one of the main founders of governance theory, clearly defines the governance in *Governance without Government, Governance in the Twenty-first*

Century and other articles: Governance is completely different from government rule. Governance is a management mechanism of a series of activities for the common goal, which still works although it is not formally authorized. Governance does not depend on the state power, but is supported by "government mechanisms, and also informal, non-governmental mechanisms" (Rosenau, 1995: 5).

In its report, *Our Global Neighborhood,* released in 1995, the Commission on Global Governance pointed out: "Governance is the sum of many ways individuals and institutions, public and private, manage their common affairs." (Commission on Global Governance, 1995: 2–3). Traditional management models emphasize control and compliance with established systems and rules. Governance emphasizes process, coordination, cross-sectoral and continuous interaction. Of course, as a public management process, governance also requires certain management rules and mechanisms.

Global governance refers to the democratic consultation and cooperation between international organizations, governments, NGOs and citizens to maximize mutual benefits. Binding international rules are formulated to address global issues such as those of environment, human rights, immigration, drugs, etc. As the core of global governance, international rules are institutional arrangements with legal liabilities to regulate international relations and international order. The effectiveness of international rules is the main indicator for examining the global governance effect. Global governance is promoted by the countries, intergovernmental international organizations, international NGOs and other external actors, which can urge the countries to follow international rules and formulate their own laws and regulations according to them.

Global governance is one of the most important theories in the international politics field after the end of the Cold War. Although the Cold War has ended, national and regional conflicts still exist extensively, and the common challenges like climate change, poverty and terrorism facing human beings emerge under the trend of multipolarization. In the context of globalization, various countries carry out increasingly frequent cooperation and exchange in politics, economy, culture, science and technology, needing to establish common rules and systems to safeguard common interests. Thus, global governance emerges, playing a positive role. A series of complex realistic problems above can not be solved by any party alone, and the theory of collaborative governance rises at the right moment, which emphasizes that governments, NGOs, enterprises and citizens should participate in public management, break the government-centered authority, and build consensus, coordination and interaction.

Climate change is a common challenge in the 21st century, which involves geopolitics, energy, economy, and development, and is one of the important factors to reshape the global pattern of politics and development. Addressing climate change requires the concerted action of the international community, so it becomes one of the priority issues of global governance. Global climate governance refers to that state actors and non-state actors carry out the climate cooperation on the basis of common values and perceptions and formulate specific governance mechanisms to address climate change and achieve sustainable development.

At the end of the 1980s, serious climate change attracted the attention of the international community. The Intergovernmental Panel on Climate Change (IPCC) established by the United Nations Environment Programme (UNEP) in 1988 is the uppermost agency for scientific assessment of climate change and has issued five assessment reports so far to jointly address climate change.

Based on the scientific findings in the assessment reports, global climate governance mechanism is being formed gradually. The UN is the main coordinator and organizer of global climate governance. The United Nations Framework Convention on Climate Change (UNFCCC) and the Kyoto Protocol (KP) are the main bases for global climate governance.

The UNFCCC was adopted at the UN Headquarters in New York in May 1992. In June 1992, it was signed by 197 countries at the UN Conference on Environment and Development held in Rio de Janeiro, Brazil, which officially came into force in 1994, becoming the core rule of global climate governance. The UNFCCC includes the principle of "common but differentiated responsibilities", which becomes the basic principle for global climate governance. Since 1995, contracting parties of the UNFCCC have convened the Conference of the Parties (COP) annually, to assess the progress of addressing climate change, having carried out negotiations on mitigation, adaptation, funding, technology and capacity building with the UNFCCC as the overall framework (see Table 1).

At COP3 of the Third UNFCCC in 1997, the KP, the first legally binding climate bill under the UNFCCC framework, was adopted, which defined legally binding quantitative emissions reduction targets, stipulating the first commitment period (2008–2012) for quantified limiting and reducing of emissions. In the KP, some institutional arrangements for global climate governance were also made. For example, any group or institution qualified to deal with the matters covered by the KP, no matter whether it is national, international, governmental or non-governmental, may attend the meeting as an observer upon the consent of the Secretariat, unless it is opposed by at least one third of the contracting parties present (Article 13.8 of the KP).

Since the 1990s, global climate governance has gradually become diverse and multi-layered under the UNFCCC framework. The UN as the main facilitator coordinates all contracting parties to engage in global climate governance on the principle of reaching consensus through consultation, ensuring that all parties have the equal right of speech in the decision-making process. Some international intergovernmental organizations, international NGOs, local organizations or institutions are invited to attend the COP as observers, to supervise the negotiation process. As people gradually realize the urgency and seriousness of addressing climate change, sub-state actors become increasingly active, becoming new forces of global climate governance.

2 Theoretical Foundation

Table 1 Summary of main COPs involved

Time	Full name	Abbreviation	Place	Main topics	Negotiation results
1997	3rd COP of the UNFCCC	COP3	Kyoto, Japan	Discussed the goals of major developed countries in reducing greenhouse gas emissions	Concluded the legally binding KP
2007	13th COP of the UNFCCC and 3rd COP of the KP	COP13	Bali, Indonesia	Discussed the emissions reduction plan after the first commitment period (2008–2012) of the KP	Adopted the Bali Road Map, confirmed the "two-track" negotiation process under the UNFCCC and the KP; Set two years of negotiation time, and decided to complete negotiations at the 15th COP of the UNFCCC and the 5th COP of the KP in Copenhagen, Denmark, in 2009
December 7–18, 2009	15th COP of the UNFCCC and the 5th COP of the KP	COP15	Copenhagen, Denmark	Discussed the emissions reduction plan after the first commitment period (2008–2012) of the KP, namely the agreement of global emissions reduction from 2012 to 2020, setting a keynote for the post-Kyoto era	Concluded the Copenhagen Accord without legal effect; maintained the principle of "common but differentiated responsibilities", Bali Road Map authorization, and dual-track negotiation process; and maximized the inclusion of all countries into the cooperative action for addressing climate change

(continued)

Table 1 (continued)

Time	Full name	Abbreviation	Place	Main topics	Negotiation results
November 29–December 10, 2010	16th COP of the UNFCCC and the 6th COP of the KP	COP16	Cancun, Mexico	Clarified the greenhouse gas emissions reduction targets of developed countries after the first commitment period of the KP expires at the end of 2012; sought consensus on developed countries' transfer of funds and technologies to developing countries to address climate change	Adopted the Cancun Agreement; and maintained the two-track negotiation process
November 28–December 9, 2011	17th COP of the UNFCCC and the 7th COP of the KP	C0P17	Durban, South Africa	Determined the quantitative emissions reduction targets of developed countries for the second commitment period of the KP; implemented the arrangements of fund and technology transferors	Established the Durban Platform for Enhanced Action, made specific arrangements for reducing greenhouse gas emissions after 2020; officially launched the Green Climate Fund
November 26–December 7, 2012	18th COP of the UNFCCC and the 8th COP of the KP	COP18	Doha, Qatar	Urged the Durban Platform to formulate the new convention for post-2020 by 2015; adopted the amendment to the KP	Adopted the Doha Agreement, and the Amendment to the KP; however, the U.S., Canada, Japan, New Zealand, Russia and other countries refused to accept, which greatly shrank the share of mandatory emissions reduction

(continued)

2 Theoretical Foundation

Table 1 (continued)

Time	Full name	Abbreviation	Place	Main topics	Negotiation results
November 11– November 22, 2013	19th COP of the UNFCCC and the 9th COP of the KP	COP19	Warsaw, Poland	Executed the negotiation tasks, consensuses and commitments mentioned in the Bali Road Map in 2007; launched Durban negotiations, and determined the goals, actions, policies and measures for addressing climate change from 2020 to 2030	Emphasized that the Durban Platform basically reflects the principle of the Convention; developed countries once again declared to support the developing countries in addressing climate change
December 1–December 15, 2014	20th COP of the UNFCCC and the 10th COP of the KP	COP20	Lima, Peru	Identified the basic elements of the 2015 Global Climate Change Agreement; enhanced emissions reduction and funding before 2020	Adopted the Lima Call for Climate Action; reached a consensus on the elements of draft agreement for 2015 Climate Conference in Paris
November 30–December 11, 2015	21st COP of the UNFCCC and the 11th COP of the KP	COP21	Paris, France	Concluded a new global agreement after the KP expires in 2020	Officially launched the post-Kyoto era
November 7– November 19, 2016	22nd COP of the UNFCCC and the 12th COP of the KP	COP22	Marrakesh, Morocco	At the first COP after the Paris Agreement officially entered into force, discussed the details about implementation of the Paris Agreement	Adopted the Marrakesh Declaration

(continued)

Table 1 (continued)

Time	Full name	Abbreviation	Place	Main topics	Negotiation results
November 6– November 18, 2017	23rd COP of the UNFCCC and the 13th COP of the KP	COP23	Bonn, Germany	Ensured to release the implementation rules for the Paris Agreement as scheduled and improve the tools and means needed to implement the Paris Agreement	Adopted the Fiji Momentum for Implementation, formed a text of balanced negotiation on various problems about implementation of the Paris Agreement, further clarified the organization form of promotion dialogue in 2018, and adopted a series of arrangements to accelerate climate actions before 2020

2.2 Two-Level Game Theory

The two-level game theory was proposed by American scholar Putnam (1988), which, in the context of global governance, provides a new theoretical analysis framework for cross-level interactive research of domestic politics and international politics, and makes different levels of synchronic analysis possible. Different from previous theories separating domestic and international politics, the two-level game theory emphasizes the simultaneous game between international and domestic politics.

A country always strives to maximize its own interests in the game process to cope with the domestic pressure and minimize adverse diplomatic consequences. Various interest groups in a country will impose various pressures to make policy makers consider their interests when formulating policies. In modern times, all countries are interdependent and value their respective sovereignty, and policy makers must take into account all factors when formulating policies. At the international occasions, negotiators of a country are responsible for negotiation with negotiators of other countries that have different interest demands; at the domestic occasions, they negotiate with various interest groups such as the government departments, private sectors and third-party institutions. Political circles pay more attention to the domestic appeals of various interest groups because they have the right to vote and will affect the election results if their interest demands cannot be satisfied in a timely manner. In other words, international negotiations should serve domestic politics. Only by understanding the transigent base line of domestic opponents can the negotiators of a country predict the results of international negotiations more accurately. The complexity of the two-level game lies in that the policy needs to be adopted domestically and as well as by other countries, while other countries also need to consider the "possibility of acceptance by their own people" (Zhong and Wang 2007). Politicians are the key actors of the two-level game, who need to consider both international and domestic needs and find a balance point between them. Politicians need to strive for maximum support at home, while shouldering the mission to ensure the smooth operation of the international mechanisms. The key of the two-level game theory is to try to obtain the largest set of international rules that can be approved at home, namely the win-set. The size of domestic win-set decides whether an international agreement can be entered into finally.

Discussing the international cooperation between different countries interdependent, the two-level game theory provides a new model, framework and perspective for analyzing the relation and interaction between domestic politics and international politics. This book adopts the theoretical framework of two-level game, mainly from the research topics and research objects perspectives.

Regarding the topics, climate change involves different fields such as environment, politics, development, economy and energy, and its scientific principles, influences and responses have multiple complexities. As a typical global issue, addressing climate change needs the participation of all human beings, which is related to the development of each sovereign state, and the evolution of the international political pattern, and is a process of two-level interaction, counterbalance and game. Both international and domestic factors need to be considered to develop appropriate strategies for spreading climate change knowledge and participating in climate governance.

Regarding the objectives, in the past 40 years of reform and opening up, China has carried out increasingly frequent and close exchanges with the world. As an emerging developing country, China plays an increasingly positive role in the global governance, maintaining international order and supporting international cooperation. According to the Report at the 19th Party Congress, China will continue to play the role as a responsible big country, actively participate in the reform and construction of the global governance system, and continuously contribute China's wisdom and strength. The stable economic growth of China is a prerequisite for China to actively carry out international cooperation. With the development of marketization, interest groups in China have gradually become stronger, having different impacts on the government's policy formulation and implementation. The Chinese government also needs to face both domestic and international challenges.

The third chapter of this book analyzes the two-level game motivation and object of China's response to climate change, and also the two-level goals of climate change communication and governance. In the study process, I revised the definitions of some concepts such as game and win-set at the domestic level according to national conditions.

3 Research Methods

This book comprehensively studies climate change communication and governance through qualitative and quantitative methods, in combination of the knowledge and theories of international relations, communication, management, sociology, psychology, and behavioral science, etc. In addition to conventional research methods such as literature review method, qualitative empirical research method, and quantitative empirical research method, the following methods are adopted.

3.1 Two-level Analytic Hierarchy Process

Analytic Hierarchy Process (AHP) is an important method for study of international relations, which assumes that the factors at one or more hierarchies will cause the occurrence of certain international events or acts. The researcher usually analyzes at a certain hierarchy according to his/her research object and purpose. A single hierarchy of analysis has significant limitations and is insufficient to explain the increasingly dependent and interactive relationship between state act and international politics. Under the theoretical framework of the two-level game, this book, through the Double AHP, analyzes the strategy change of China's three major stakeholders in climate change communication and governance at the international and domestic levels.

3.2 Analysis of Stakeholders

The stakeholder theory originates from business management, and a foundation work about it is *Strategic Management: A Stakeholder Approach* by R. Edward Freeman published in 1984. According to Freeman, the stakeholder refers to "a single person or a group of people who can be influenced by, or influence, the failure or success of an organization" (Freeman, 2006). Freeman proposed the stakeholder theory in the context of globalization. In the early 1980s, enterprise production factors and product markets were rapidly globalized, and the enterprises faced fierce competition from home and abroad and also new uncertainties. Original shareholder supremacy has lost its advantage in the new environment, and the introduction of stakeholders can make the enterprises better understand the external environments and changes and achieve effective organization and management. Stakeholder theory involves three core concepts, namely three stages of development.

First, Stakeholder Influence, emphasizes the interaction between an organization and its stakeholders, especially the impact of stakeholders on an organization's strategy and performance. The study at this stage show the tendency of instrumentalism and organizational standard, and stakeholders are regarded as external environmental factors or management objects, but the research perspectives are relatively simple.

Second, Stakeholder Participation, stresses that stakeholders "use a set of procedures to exert influence on and share control over the decisions, activities and resources that influence them" (World Bank, 1996). At this stage, the stakeholder issues are included in the scope of organization, and dual perspectives replace the single perspective.

Third, Stakeholder Co-governance, means that the company/organization and all its stakeholders mutually balance and co-govern. At this stage, stakeholders are treated as subjects equal to companies and organizations. The transformation of stakeholders from management objects to governance subjects is a major symbol of in-depth development of stakeholder theory.

The above three stages are inclusive and progressive from the outside to the inside, which show the evolution and leap of stakeholder theory (see Fig. 1).

The stakeholder analysis method adopted in this book emphasizes the third stage, namely Stakeholder Co-governance. As the global governance model enjoys popular support in the early 21st century, Stakeholder Co-governance with the influence far beyond the scope of corporate governance, has changed from a theoretical paradigm to a practical model, and has been widely used in the field of social life.

Breaking the original single governance model, the stakeholder theory is applied in the related fields such as risk control and public policy, which can help preventing and controlling the defects in the policy-making process. For the stakeholder analysis method, the key is to define the stakeholders for the stakeholder co-governance.

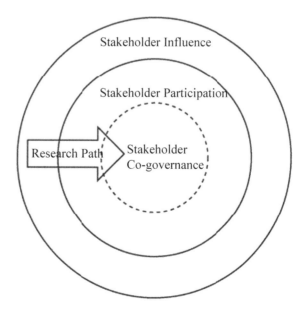

Fig. 1 Three development stages of stakeholder theory. *Source* Shenyu Wang (2008)

Stakeholder analysis method is an important tool for understanding social and institutional contexts, who are affected (positive or negative), who will influence the development (positive or negative), what individuals, groups or organizations need to be considered, and how to exert different influences.

From the perspective of communication, the boundary between the communicator and the receiver has been broken. In the era of mass communication, communication technology was mastered by a small number of people or organizations, and the receivers were passive and inferior. In the era of new media, the pattern of monopoly by traditional mass media has been broken, and the general public as the traditional receivers have owned more abilities to generate, spread and receive information, thus gaining the abilities to express opinions, exert influences and participate in decision-making and organizational evaluation. Now, everyone is a communicator and also a receiver. The traditional theory of communication and receiving is no longer applicable to the current need of climate change communication and governance study. The subjects of involvement in communication are all independent actors that are closely related.

From the perspective of global climate governance, with the advent of globalization era, unilateral actions can no longer solve all problems, and co-governance through joint actions of multiple actors becomes an irresistible trend. Climate change is one of the top issues concerning global governance. For global climate governance, stakeholders such as sovereign states and non-state actors are encouraged to follow common values and carry out transnational cooperation in the climate field. Relevant parties involved in climate governance will actively or passively participate in climate change communication.

Both the two-level game theory and the stakeholder theory are generated in the context of globalization. The two-level game theory emphasizes the game at two levels, which is the analytical framework of this book; the stakeholder theory emphasizes the maximization of all parties' interests. Through stakeholder analysis, basic research objects are identified.

3.3 Tracking Analysis

Tracking analysis is a method of sociological research, which means the periodic investigation of the same group of subjects at multiple time points in a relatively long period. Compared with the one-time, single-point cross-sectional studies, tracking analysis can explain the time sequence of two kinds of social phenomena with causal relationships.

The result of the tracking analysis from 2009 to 2015 is summarized for the fifth chapter of this book. First, in the past six years, China has completed the transition from passive follow-up to active leadership, successfully reversing the image of China in global climate governance. This is the first time that China masters the right to formulate rules and the right to speak about global governance at the fastest speed in the past 40 years of reform and opening up. China's experience in participation in global climate governance is of reference significance for China to address other global governance issues and implement the "Belt and Road" Initiative.

Second, the KP came into effect in 2005. According to the Bali Road Map adopted after the climate negotiations in Bali, Indonesia in 2007, if a new agreement cannot be concluded in 2009, global greenhouse gas emissions will not be limited any longer when the first commitment period of the KP expires in 2012. The second commitment period of the KP is from 2013 to December 31, 2020. At the end of 2015, a new global agreement of the Paris Climate Conference should be concluded to substitute the KP. In short, the 2009 Copenhagen Climate Talks pushed the discussion of the KP to a new height, which focused on how to address climate change after 2012. 2015 is the first year of "post-Kyoto era", bidding farewell to the old era of global climate governance.

Seen from the perspective of communication effect, the 2009 UN Climate Change Conference in Copenhagen, at which although no legally binding agreement was concluded as scheduled, aroused continuous coverage of global media because of its importance, which made climate change and the UN Climate Change Conference gain unprecedented attention. In 2015, China and the U.S. jointly promoted the conclusion of the Paris Agreement, which became a hot event that year. In 2009 and 2015, climate change communication reached two peaks.

4 Research Framework and Value

4.1 Research Framework

The book composed of eight chapters analyzes three major stakeholders from both international and domestic perspectives, and the strategy change of climate change communication and governance from 2009 to 2015 (Chaps. 2–5), carrying out a new round of empirical research according to the latest progress from 2015 to 2018 (Chaps. 6–8) and summarizing the experience in a timely manner, which is of reference significance for China's participation in the global climate governance, etc.

This chapter, introduces the research background and defines climate change communication as a strategic tool for climate governance, pointing out the purpose of communication is to promote more effective governance, elaborating the theoretical basis and research methods, and confirming the research framework.

Chapter 2 **Overview of Researches on Climate Change and Climate Change Communication**, clarifies the concept of climate change on the basis of reviewing the relevant literatures at home and abroad, defines the uncertainty as a normal state of scientific research, and responds to some existing confusion about "whether climate change really exists", distinguishing climate change communication from environmental communication, development communication, health communication, science communication, risk communication and political communication through overview of international and domestic researches on climate change communication, and clarifying the definition and research value of climate change communication.

Chapter 3 **Two-level Game: Analytical Framework for Climate Change Communication and Governance in China**, analyzes the two-level game motivation and object of China's response to climate change, revising the definitions of some concepts such as 'win-set' according to national conditions.

Chapter 4 **Three Major Stakeholders: Basic Units of Climate Change Communication and Governance in China**, revises three indicators (legality, power and urgency) on the basis of Mitchell stakeholder scoring method, and adds correlated indicators, analyzing China's stakeholders of climate change communication and governance from both international and domestic perspectives, and defining the government department, media, and NGOs as key stakeholders.

Chapter 5 **Tracking Analysis of Role and Strategy Change of Governments, Media and NGOs (2009–2015)**, assesses the performance of three major stakeholders in these six years, defines their roles, and introduces the strategy change in communication and governance through empirical analysis, summarizing the main findings, and testing the applicability of the two-layer, multi-dimensional research space in the field of climate change communication and governance.

Chapter 6 **China's Climate Change Communication and Governance in the Post-Paris Era**, summarizes the challenges China faces after the 2015 UN Climate Change Conference in Paris, analyzes the coping strategy for the post-Paris era, and identifies realistic challenges, proposing three strategic suggestions from the perspective of trend analysis.

Chapter 7 **New Stage of "Dual Transition" in Global Climate Governance and China's Choice**, analyzes the new situation of global climate governance from 2016 to 2017, and China's strategic choices in a two-layer, multi-dimensional research space on the basis of making clear macroeconomic governance situation.

Chapter 8 **Conclusion and Prospect**, summarizes the value of two-level, multi-dimensional research space and comprehensive application of relevant theoretical tools on the basis of case analysis and follow-up study, and discusses the main conclusions drawn after verification of this research framework (namely China's path of climate change communication and governance), making outlook on China's future participation in global governance.

4.2 Research Value

Currently, China has the will and ability to further engage in global governance, make contributions to the international community and actively implement the "Belt and Road" Initiative. Summarizing China's successful experience in engagement in global climate governance and studying the feasibility of experience spillover can enrich China's solutions and help China better make innovative contributions to the global governance.

First, this book, based on relevant theories and long-term study results of climate change communication and governance in China, constructs a two-level, multidimensional research space, and summarizes China's path of climate change communication and governance, which is comprehensive and supplementary to the diverse and multi-layered global climate governance research. New research space makes it possible to comprehensively comb China's successful experience in engaging in global climate governance.

Second, the two-level game theory involving international relations can more accurately explain the decision-making plight of a single country in participating in global climate governance. The two-level game theory is a research branch of rational institutionalism, emphasizing the domestic and international interaction. Chinese scholars have made many useful evaluation-oriented and empirical research attempts in the aspects including China's diplomatic practice, foreign economic policy, and business diplomacy practice. This book makes a beneficial attempt in applying the two-level game theory to the research on climate change communication and governance.

This book, through the stakeholder analysis method of management and the tracking analysis method of sociology, makes the interdisciplinary, three-dimensional and cross-over research on study objects, constructs a two-level, multi-dimensional and panoramic research space, tries breaking through the existing research methods on the single communication standard, expands the theoretical depth and research level, and promotes the interdisciplinary perspective integration. Based on many years of my experience in following up the UN Climate Change process, I track and study the change of climate change communication strategies of governments, media and NGOs using the two-level analysis method, define the definition and value of climate change communication in the context of global climate governance, and broaden the research space of climate change communication, enriching the research content of global climate governance.

Third, for better global research in the field, international NGOs and domestic social organizations in China deserve more attention. As the researches are deepened, the role of international NGOs in global environment and climate governance has been highlighted, and domestic civil society organizations have gradually attracted the interest of researchers. In particular, NGOs, relying on their solid network and mass bases, have become one of the main forces of implementing the "Belt and Road" Initiative, having great potentials in promoting the interlinked popular feelings. A correct understanding of the types, functions and influences of NGOs can help China better communicate with the world. With a wealth of experience of work in media, international NGOs and intergovernmental organizations, I demonstrate the interactive relationship between governments, media and NGOs when reviewing their strategy changes, hoping to provide useful reference for follow-up researches.

Fourth, the premise of studying climate change communication and governance is to reach a consensus on the basic issues including the authenticity of climate change, the value and positioning of climate change communication research, etc. Over the past 10 years, I've encountered the most divergences in this regard. However, lack of consensus will affect the depth of discussion. This book explains the uncertainty of climate change through analysis and comparison of existing literature at home and abroad, expounds the importance of precautionary principle, and distinguishes climate change communication from environmental communication, political communication and other related application communication fields, positioning climate change communication as an independent research direction, which lays a theoretical foundation for the cross-over study and other studies by the readers interested in climate change communication. Response to the doubt about the uncertainty of climate change through literature review is of practical significance for reaching the interdisciplinary action consensus.

In short, the theme of this book is in line with the major social concerns of China, having certain practical significance. This book makes an applied research, making the case analysis based on existing theories. Currently, empirical researches and interdisciplinary research attempts are relatively realistic. As the author, I make an empirical research according to my long-term observation results and in-depth interview materials, instead of making breakthrough and innovation in interdisciplinary theories, and try applying the relevant theories and methods concerning international relations, public administration, sociology and communication for the research on climate change communication and governance, to build a comprehensive research space. I hope my explorations can play a role in broadening the research horizon of climate change communication and governance.

Chapter 2
Climate Change and Climate Change Communication

1 Summary of Climate Change Research

1.1 Certainty of Climate Change

By time, climate change research is divided into paleoclimatology research and modern climate change research. Closely related to paleogeology, paleontology, geochemistry, and atmospheric physics, paleoclimatology mainly studies the formation of climate during the geologic epoch. Dated back to the late nineteenth century, modern climate change research refers to the climate change research based on systematic data of observation by meteorological instruments. Modern climate change is different from paleoclimate change mainly due to impacts of human factors. In the 1980s, modern climate change research was not limited to the scope of natural sciences any longer, but was combined with social sciences such as politics, society, economy, sustainable development, and international relations, and was further promoted to the level of global climate governance.

In modern times, there are two main definitions of climate change. Intergovernmental Panel on Climate Change (IPCC) defines climate change as any change in climate status over time, emphasizing the impact of human activities in addition to the natural variability of climate change. The United Nations Framework Convention on Climate Change (UNFCCC) also emphasizes natural variability, holding that climate change is caused by direct or indirect human activities, and global warming is attributed to developed countries' consumption of massive fossil fuels and emission of excessive carbon dioxide and other greenhouse gases during industrialization.

Discovery of the greenhouse effect is a key node for modern climate change research. Frenchman Joseph Fourier (March 21, 1768–May 16, 1830) is widely recognized as the first scientist finding out the principle of greenhouse effect. He said that the Earth can preserve heat mainly because the atmospheric layer re-reflects the heat from the Earth back to the ground. In 1895, Swedish scientist Svante

August Arrhenius (February 19, 1859–October 2, 1927) developed the first theoretical model used to calculate the impact of carbon dioxide on the Earth's temperature, and first proposed the concept of greenhouse effect. In 1961, U.S. scientist Charles David Keeling (April 20, 1928–June 20, 2005) proposed the "Keeling Curve", which provided a long-term data support for the concentration increase of carbon dioxide in the atmosphere and laid a foundation for global climate change research. In the following 20 years, signs and analyses of global warming have been accumulated continuously. At the First World Climate Conference held in 1979, representatives of participating countries basically reached a consensus on the fact and possible impacts of climate change, kicking off global climate governance. In 1988, IPCC was formally established with the promotion of the United Nations Environment Programme (UNEP) and the World Meteorological Organization (WMO).

Open to all member states of the UN and the WMO, IPCC shoulders the mission to collect and collate existing global information on climate change science for the comprehensive and objective assessment. To date, the IPCC has issued five assessment reports, each of which plays a key role in promoting global climate governance.

In 1990, IPCC released the first assessment report. According to the data of global mean surface temperatures in nearly 150 years summarized by the four most representative research institutions in the world, global mean temperature has shown an upward trend despite the fluctuation of "warm-cold-warm". The first assessment report cited this finding, verifying that climate change is really occurring. This assessment report directly promoted the conclusion of the UNFCCC at the United Nations Conference on Environment and Development in 1992. UNFCCC is the first global convention for the international community to address climate change and also the basic framework for global climate governance. Since then, the formulation of international climate mechanisms has been embarked on the right track.

In 1995, the IPCC released its second assessment report, which clarified that human activity is the main cause for greenhouse gas growth. Based on this report, Kyoto Protocol, the first international agreement on control of greenhouse gas emissions, was entered into in December 1997.

In 2001, the IPCC released its third assessment report, which provided a clearer standard for international emissions reduction, linked climate change with sustainable development of human, and contributed to the successful signing and entry into force of the Kyoto Protocol in 2005.

In 2007, the IPCC released its fourth assessment report, which made a comprehensive assessment of the fact, causes and impacts of climate change from different angles, and directly promoted the formulation of the Bali Road Map at the end of 2007, officially launching the dual-track negotiations.

In 2014, the IPCC issued its fifth assessment report, which was jointly drafted by more than 800 scientists unprecedentedly, and also was most comprehensive. The report confirms that climate change is occurring all over the world and global warming is undoubted. The concentration of greenhouse gas (mainly including carbon dioxide, methane and nitrous oxide) in the atmosphere has risen to an unprecedented level over the past 800,000 years. This assessment report further affirms that greenhouse gas emissions and other anthropogenic factors are "very likely" to directly cause the climate change observed since the mid-twentieth century. According to the report, gradual climate warming may increase the likelihood of serious, widespread and irreversible impacts, including severe and widespread impacts on unique and threatened systems, mass extinction, huge risks on global and regional food security, as well as high temperature and high humidity threatening human's normal activities (including but not limited to food planting or outdoor work in some areas in part of a year).

The IPCC will complete drafting of the sixth assessment report composed of three working group reports by 2021. The comprehensive report will be completed in accordance with certain procedures by 2022 and will be delivered to the UNFCCC in a timely manner. By then, each country will review their progress in achieving the global warming goal of below 2 °C, and try to control it to below 1.5 °C.

1.2 Uncertainty of Climate Change

Due to the limited level of scientific cognition, human's understanding of climate change is still not comprehensive, and scientific circles are still controversial about global warming extent and causes, predictions by climate models, and impacts on humans.

In his article "Uncertainty and Climate Change", U.S. statistician L. Mark Berliner said that the inaccuracy of existing observation data and climate models brings uncertainties to climate science research. In the article entitled "Cognition of Certain Uncertainties in Climate Change Research", Quansheng Ge et al. summarized several uncertainties in climate change research, discussing about the key areas worthy of special attention, and pointing out: humans' scientific understanding of climate change is still inadequate with some uncertainties, so the decisions on climate change are not necessarily completely rational. In the article "Summary of Global Climate Change Controversies", Xiaoyu Zhang and Wenjing Shi summarized the current controversies of the scientific circles over the global climate change and related issues, including the climate change trend and extent, causes, climate model prediction, impact on humans, and feasibility response policy.

Professor Geoffrey Heal of Columbia University believes that climate change is complex, non-linear and dynamic, with uncertainties. Even if an economic model can analyze the situation after the trend of climate change is prevented, it can hardly simulate the extent of climate change reversing through control of greenhouse gas emissions. Moreover, climate change will bring about a series of consequences such as disappearance of species. Even if climate change can be controlled, disappeared species can no longer appear, which is beyond the prediction scope of economic model.

Quansheng Ge and other scholars think that the climate change is very complicated, and human beings cannot have a clear scientific cognition of climate change in a short period of time. Moreover, climate change involves different disciplines and areas, thus inevitably resulting in cognitive differences. Shaowu Wang also stresses that climate change cannot be studied thoroughly forever, because new problems will arise one after another.

In addition, others think that conflicts of interests may also affect the objectivity, accuracy and fairness of scientific activities. In the article "Conflicts of Interests in Science Communication of Climate Change", Yong Luo and Yun Gao specially pointed out: climate change uncertainty is emphasized and exaggerated mainly because individuals, organizations, institutions and climate change certainty supporters have conflicts of interests in political stand, economic interests, academic views, or media publicity.

Despite certain uncertainties in climate change research, consensuses have been reached in many aspects, including the global warming trend, which will seriously affect humans and ecosystems although the extent of warming is still uncertain. Among many causes for global warming, excessive emission of greenhouse gases is most probably the main factor.

1.3 Precautionary Principle

Like traffic prediction and meteorological prediction, climate prediction also has uncertainties. Each assessment report issued by the IPCC contains a description of uncertainty in the conclusion part, in which some conclusions based on enough evidences are highly credible, while some conclusions have medium credibility due to less evidence support. In this context, the IPCC specially developed a set of "confidence terms" to accurately describe the different accuracy of predictions. Uncertainty is the norm of scientific research but not the unique feature of climate research. Researchers at Columbia University's Center for Research on Environmental Decisions emphasize: "Scientists can never accurately predict climate change but can only make predictions based on the best available data so as to quantify the uncertainties." (Center for Research on Environmental Decisions, 2009)

Although most people are psychologically inclined to accept certainty and "security sense", rather than uncertainty and "sense of out of control". Uncertainty has become a new normal of today's risk society rapidly developing, so that people need to face different uncertainties and potential impacts with a more positive mindset.

A large number of literatures show that different disciplines have made different degrees of researches on uncertainty. As early in 1921, U.S. economist Frank. H. Knight (November 7, 1885–April 15, 1972) distinguished uncertainty from risk in his book *Risk, Uncertainty and Profit*. He divided uncertainty into the uncertainty with probability distribution and the uncertainty without objective probability distribution. Knight thought that the latter uncertainty is the real uncertainty. From the perspective of management, Milliken defined uncertainty as "the failure to accurately predict the state of the environment due to lack of information or inability to distinguish relevant and unrelated data." (Milliken, 1987: 133–143) In the work "Uncertainty in Integrated Assessment Modelling: From Positivism to Pluralism", Van Asselt and Rotmans held that too much information available might also bring about uncertainty. In the article "Defining Uncertainty: A Conceptual Basis for Uncertainty Management in Model-Based Decision Support", Warren Walker et al. tried matrix analysis of uncertainty from the perspectives of location, level and nature to provide policy makers with a basis for uncertainty management. Among them, location refers to the occurrence of uncertainty in the complex model; level refers to the position of uncertainty between two limits, namely "certainty" and "complete ignorance"; nature of uncertainty refers to the incompleteness of knowledge or the inherent variability of the phenomenon. In the article "The Precautionary Principle and the Uncertainty Paradox", Marjolein B. A. et al. put forward the modern uncertainty paradox, i.e., people realize that science cannot provide hard evidences for uncertain risks, but increasingly hope to look for certain and decisive evidences relying on science.

The cognitive background of this paradox is that people have been accustomed to the "traditional and positive" role of revealing the truth which knowledge and science play. But now, situation is changed. The ubiquitous uncertainty risks pose challenges to conventional scientific analysis, and people need to rethink the value and role of science from the interdisciplinary perspective and redesign alternative procedures.

In addition, UNESCO has made two supplements to "uncertainty" in its report: "high-quality science does not require lower uncertainty"; "for the issues that have very high uncertainty degrees or knowledge gaps, or whose decisions involve large benefits, those uncertainties that are difficult to quantify may be much more important than those quantifiable." (UNESCO, 2005)

In his doctoral thesis "Certainty in Uncertainty: Uncertainty Management in Risk Assessment of Anthropogenic Climate Change", Van der Sluis said that newly acquired knowledge may reveal more uncertainty instead of eliminating uncertainty as expected. Moreover, such uncertainty has existed before emergence of new knowledge, but is only ignored or underestimated. Therefore, more knowledge may make people's understanding more limited, or make the process of problem handling more complicated than that previously imagined.

Researches on anthropogenic climate change involve multiple uncertainties that cannot be eliminated all. In this case, the classic model of scientific analysis, namely the way of solving the puzzle under an uncontroversial framework will not work. No matter how successful this model was in single-disciplinary research, it seems powerless for the interdisciplinary issues related to transnational and intergenerational risks.

Emphasizing and recognizing the uncertainty of science is not an excuse for "inaction" or "maintaining the status quo". Addressing uncertainties through new models has become an active or passive consensus of all sectors of society. Precautionary Principle which rose in the 1970s is researched and applied most widely.

The Precautionary Principle as a rational alternative solution is applied when science fails to provide a clear answer during risk assessment and management, including the wisdom of action in an uncertain state (UNESCO, 2005). Application conditions of the Precautionary Principle include complex natural system and social system that determine the causal relationship between any human activity and its consequence, as well as scientific uncertainty difficult to be quantified when hazards and risks are described and assessed.

In the report "Precautionary Principle 1896–2000" released in 2001, European Environment Agency listed 12 cases including those of asbestos, fungicides, water pollution in the Great Lakes region, and epidemics from 1896 to 2000, and illustrated the catastrophic and irreparable consequences from failure to take preventive measures in a timely manner. It is worthwhile to take preemptive preventive measures to avoid the possible consequence, damage and loss.

Over the past 30 years, the Precautionary Principle has become a basic principle of international treaties and declarations about sustainable development, environmental protection, health, trade and food security. The UNFCCC also stipulates: "All contracting parties should take Precautionary measures to prevent or minimize the climate change and mitigate adverse impacts. In case of any risk of causing serious or irreversible damages, such preventive measures should not be postponed on the ground of no complete certainty in science." (UNFCCC, 1992).

As the Precautionary Principle played an increasingly important role in decision-making, a debate on whether it was a political trick or a scientific act was aroused by Western academic circles in the late twentieth century. Gray (1990), Stebbing (1992), Bewers (1995), Dovershe Handmer (1995) and other scholars successively published articles, holding that the Precautionary Principle is only a political management philosophy, and questioning its scientific nature. Later, different scholars responded to the question about the scientific nature. Among them, Bernard D. Goldstein, a well-known U.S. scholar in public health, published the most influential article "Precautionary Principle and Scientific Research Are Not Antithetical" in 1999.

Bernard thought: First of all, the issues about the Precautionary Principle were originally based on scientific research; second, the application of the Precautionary Principle will stimulate further scientific research so as to identify the real problems; third, research agendas should be set for responsible preventive actions, to ensure that the actions are reasonable and effective; finally, any intervention on the Precautionary Principle is subject to the supervision and evaluation by the stakeholders including the public, rather than only "insiders", so as to judge whether the intervention matches with the target and ensure the justice of intervention. Bernard's views are widely recognized, and the Precautionary Principle as a scientific tool has been applied in related fields.

"No-regret action" corresponds to the Precautionary Principle, that is, no matter whether preventive measures are necessary, they are beneficial to economic development. The key of "no-regretaction" lies in that for some risky problems, we will not regret after taking preventive actions even if no problem happens but will regret when some problems occur if you don't take action. In the "Precautionary Principle" report, UNESCO raised this principle to the legal level: "Precautionary Principle is legally significant, which cannot be ignored by each country, its legislators, policy makers, and courts. Since the Precautionary Principle was recognized as an element of international law, it has become part of the general principles of environmental law and owned irrefutable legitimacy in guiding the interpretation and application of all existing legal norms." (UNESCO, 2005).

In short, uncertainty of climate science under the interdisciplinary and globalization background has become a new normal. In today's complex risk society, people should not avoid or deny uncertainty, but should adjust the dependence on certainty under the original single scientific framework, boldly face the uncertainty of climate change, and actively seek effective intervention paths under the guidance of the Precautionary Principle.

2 Summary of Climate Change Communication Research

2.1 Definition of Climate Change Communication

No matter whether you recognize the severity of climate change and its serious impacts on humans and ecosystems, climate change has undoubtedly become "one of the most serious challenges facing humans in the twenty-first century" (Schneider, 2011: 53–62).

Since the 1970s, climate change research has been expanded from pure environmental area to comprehensive areas of politics, economics, development, and environment, and has aroused a global action to address climate change together proceeding from pure research in laboratories. Such evolution is directly related to the increasing importance of climate change issues and the increasing urgency of global response to climate change. Climate change communication has played a key role in this process.

Since 2000, seminars and special studies on climate change communication have been rising. With the support of the Canadian Climate Change Action Fund, the Canadian Environment Assessment Agency entrusted the University of Waterloo to hold the International Symposium on Climate Change Communication from June 22 to June 24, 2000, which 250 representatives invited from Canadian governments, universities, NGOs, independent consulting firms, local communities and media attended. The representatives present thought due to complexity, climate change could not be directly understood by the audience, thus bringing certain difficulties to climate change communication. In addition, greenhouse gas emissions reduction target was not made clear, and the stability of the climate model needed to be observed at that time, so that climate change communication became more difficult. The Symposium also put forward three expectations for climate change communication, namely, enhancing awareness, deepening understanding and inspiring action.

The sixth chapter of UNFCCC highlights the necessity and importance of climate change communication and stakeholder participation in addressing climate change: "To address climate change, immediate actions must be taken to carry out publicity and education in local communities and enhance public awareness of foreseeable risks. As an important part of lifelong learning, community-based climate change communication and education is of important significance to enhance awareness, build partnerships, influence behaviors and promote public participation in sustainable development."

Because the climate change communication rises not long ago, academic circles have not yet reached a consensus on its definition but paid more attention to applied research. In 2007, EU Spatial Planning defined the climate change communication strategy: climate change communication aims to "make more people aware of climate change, raise public awareness of climate change crises, increase public responsibility to adapt to and mitigate the impacts of climate change, and provide the best practice recommendations and cases about climate change and emissions reduction" (EU Spatial Planning, 2007). Swedish scholar Victoria Wibeck concluded from the literature that the ultimate goal of climate change communication is to reduce the impact of climate change and achieve sustainable development through public participation (Victoria Wibeck, 2014). In his book *Climate Change Communication Theory and Practice*, Chinese scholar Baowei Zheng said: "Climate change communication is to address climate change issues by spreading the climate change information and related scientific knowledge to the public so as to change their attitudes and behaviors. In short, climate change communication is a social communication campaign on climate change information and knowledge to ultimately address climate change issues" (Zheng, Wang, & Li, 2011).

Based on previous studies and 10 years of practical experience participation in global climate governance, I make such a conclusion: "Climate change communication is a joint communication campaign by stakeholders at different levels of global climate governance to promote more effective governance. It is a strategic tool for climate governance."

2.2 Relationship Between Climate Change Communication and Six Major Application Communication Modes

What is the difference between climate change communication and other communication modes such as environmental communication, risk communication, health communication, political communication and economic communication? Can the accumulated experience in these aspects be directly applied to climate change communication? Is it necessary to give special academic attention to climate change communication?

In combination of practical problems, U.S. climate change communication researcher Susanne Moser combed the root causes of climate change, including the insufficient visibility and instantaneity of climate change, the lack of sense of achievement due to delay of the climate system, the cognitive limitation despite technological progress, the complexity and uncertainty of climate change, the insufficient signal to make changes, and the egoism. "These causes and the complexities of interaction between humans and climate make climate change communication more challenging than communicating environment, risk or health issues." (Moser, 2010).

In addition to the causes emphasized by Moser, relevant applied communication modes such as environmental communication, development communication, health communication, science communication, risk communication and political communication, as well as their relationship with climate change communication, were combed here to better understand climate change communication.

1. Environmental communication and climate change communication

The earliest research on environmental communication stared in the 1980s, mainly being concentrated in the U.S. Robert Cox (November 28, 1917–June 22, 2013) (Cox, 2006) defined environmental communication as a means of understanding the environment and the relationship between human and natural environment. Constructing of environmental issues by this means aims to achieve communication between human and environment.

Climate change was defined as an environmental issue by Western scholars in the early period. For a long time, communication of climate change information became one of the research contents of environmental communication. In recent years, along with the confirmation of anthropogenic factors of climate change and the worldwide attention to climate change, interdisciplinary nature of climate change has been gradually recognized, and more and more specialized researches on climate change communication have been conducted. However, climate change communication originated from environmental communication, and many Western scholars and institutions studying climate change communication also have the background of environmental communication. For example, Yale Center for Climate Change Communication (YPCCC) is subordinate to Yale School of Forestry and Environmental Studies. Now climate change communication research still absorbs nutrients from environmental communication. The research fields of environmental communication have been extended to "natural representation of environmental discourse and rhetoric, media and environmental news, public participation in environmental decision-making, social marketing and environmental campaign, environmental cooperation and conflict resolution, risk communication, mass culture, and green market." (Xu, 2013). Inspired by the latest progress in environmental communication research, or using environmental communication methods for reference, many researchers have been contributing to the climate change communication research.

2. Development communication and climate change communication

In the 1950s, the theory of development communication emerged in the U.S. Development communication aimed to educate and influence the public through media and stimulate public participation so as to strategically promote social development. Development communication focuses on systematic intervention in social processes through the knowledge communication.

Climate change is ultimately an issue of development. Some theories of development communication have been applied in the fields of environmental communication and climate change communication, including the most typical theory of participatory communication, which focuses on the individual role in the process of mass communication to express the individual's own opinion and to be heard. The public is encouraged to participate in decision-making discussions related to their own interests. In recent years, public participation in environmental decision-making has become a hot topic, and the public participation in climate change communication has received increasing attention.

The theory of entertainment education is also of reference significance to climate change communication. The theory emphasizes the entertainment orientation of communication methods and stimulates the public interest in the content of communication through the entertainment elements. In recent years, the theory of entertainment education has been applied in related fields such as health communication and environmental communication. For instance, with the slogan of "soap operas can also change the world", the Population Media Center in the U.S. aims to change people's behaviors through entertainment education so as to improve people's health and well-being.

3. Health communication and climate change communication

U.S. communication scholar Rogers Ernest Malcolm Whitaker (January 15, 1900–May 11, 1981) thought that health communication means introduction of medical research results to the public, so as to enhance public awareness of health and bring about changes in behavior. Changes in public attitudes and behaviors can reduce prevalence and mortality rates and improve life quality and health level. In the new media era, public communication falls in a dilemma of communication content fragmentation, ubiquitous communication, and social media mainstreaming. Chinese scholar (Hu, 2012) said that communication in the traditional sense has one-way and linear limitations, which were not particularly prominent in the era in which traditional media played a dominant role. However, after the emergence of new media, such limitations are magnified. The traditional communication mechanism featuring infusion malfunctions, and health communication needs to adapt to the new multipoint interlacing characteristics of online communication. Baijing Hu emphasizes that not only health communication, but also public communication in political, economic, social and cultural fields faces such challenges.

Both climate change communication and health communication ultimately aim to solve related problems by promoting actions and face the real challenge to innovate traditional communication modes. The methods explored in reality can be mutually used for reference. In addition, research models, experience and lessons of health communication around innovation diffusion, social marketing and social learning are of reference significance to climate change communication researchers.

In addition, one of the obstacles to climate change communication is that climate change is very "distant" to people's life. However, researches gradually show that climate change has various impacts on human health. Scholar Victoria said: "Climate change communication under the health framework can better motivate the public to take a positive response" (Victoria, 2014).

4. Science communication and climate change communication

In the book *Social Function of Science* published in 1939, U.K. physicist J. D. Bernard specially discussed science communication, thinking that science communication involves communication among scientists, science education and science popularization, aiming to spread the scientific knowledge. Chinese scholar Huajie Liu holds that science communication is divided into two levels, and values Level-2 highly. Level-1 is the top-down spread of specific scientific knowledge which is always unconditionally beneficial to society. Popularization of science in the traditional sense belongs to the scope of Level-1 Level-2 weakens the attention to knowledge, emphasizing the impact of science on society. He highlights that science communication is a dynamic mesh feedback system involving multiple behavior subjects.

Like the science communication, climate change communication also experienced the process from traditional popularization of science to multi-communication. Here, the dynamic feedback system of climate change communication is also a focus of this book. We can say that climate change communication and science communication are overlapped somewhat. However, from the perspective of content and mechanism, science communication focuses on the sharing and popularization of scientific knowledge. Although climate change belongs to the scope of scientific knowledge, climate change communication aims to solve problems by raising awareness and promoting action, so as to improve governance. Therefore, climate change communication and science communication cannot be simply equated.

5. Risk communication and climate change communication

Risk communication originated from sociology and was especially influenced by German sociologist Ulrich Beck. In 1986, Beck first used the concept of "risk society" to describe the society in the post-industrial period. The material wealth of post-industrial society is much richer than ever before, but brings risks to human beings in many areas including ecological environment, economy, military, etc. The ecological environment risk first attracted social concern.

"Risk" and "danger" are not equivalent. "Danger is real, but risk is a social construction." (Solvic, 1997). Covello, Zimmermann, Kasperson, Palmlund and other scholars (1986) pointed out that good risk communication should have the functions of enlightenment, right to know, attitude change, legitimacy, risk reduction, behavior change, public involvement, and participation. Risk communication can further promote mutual understanding and clearer definition of risks. Realizing the existence of risks, one can take the initiative and positive attitude so as to change the passive situation. By adopting the risk communication mechanism, we can reduce risks and take protective actions.

Grabill and Simmons et al. (1998) summarized three kinds of risk communication modes, namely science and technology, negotiation and criticism. The criticism orientation was proposed after realizing the limitations of the first two orientations. In different modernization scenarios, risk communication emphasizes the construction characteristics of risks.

Xie (2012) combed the history and current situation of risk communication researches in China and abroad, believing that risk communication mainly functions to solve the consequences of risks in different areas on the basis of risk analysis. Risk communication has been applied for interdisciplinary researches, rather than only single disciplines. The theory of risk communication has also been put into practice and developed.

Climate change is a typical risk, so risk communication is also optional for climate change communication. Risk communication for climate change communication can help the public better understand the certainty and uncertainty of climate change risks, the related response principles, and the effect of proactive response. Cognition is the premise of action, and cognition of risks is conductive to promoting participation and action. Of course, in addition to risk communication, frameworks of health, politics, environment, development and science are all optional for climate change communication.

6. Political communication and climate change communication

In the book *An Introduction to Political Communication,* Brian McNair defined political communication as any purposive politics-related communication. The research fields of political communication include political information, news media, public opinion and new media. The basic theories include political rhetoric theory, agenda setting theory, Spiral of Silence, persuasion theory, etc. Climate change involves international and Chinese politics, and international relation is an important perspective for considering global climate governance. However, if we understand climate change completely from the perspective of political science, we are very likely to fall into the mire of conspiracy theory, while ignoring the scientific nature of climate change.

The results of reviewing and comparing the above-mentioned researches on applied communication (see Table 1) are mainly as follows.

First of all, climate change communication is related to but different from six other communication modes in orientation. Climate change communication research started late, absorbing the experience in relevant applied communication researches. Just as climate change research can be conducted from different perspectives, climate change cannot be simply understood as a single discipline issue. Different communication modes have different characteristics, so it is necessary to give special academic care to climate change communication. Simple establishment of subordination relation will easily obliterate the development potential.

Second, the above-mentioned communication modes as different kinds of applied communication are all not clearly defined in concept and are different in development degree. Their theoretical frameworks are still under construction, and their methodologies are still to be studied, so that an independent discipline has not been

Table 1 Application directions of six climate change communication modes

Communication mode	Time of origin	Place of origin	Research content	Main theory
Environmental communication	1970s–1980s	U.S.	Natural representation of environmental discourse and rhetoric, public participation in environmental decision-making, environmental cooperation and conflict resolution, risk communication, mass culture, and green market, media and environmental news, social marketing and environmental campaign	Theories related to meaning interpretation, semiotics, discourse analysis and communication
Development communication	1950s	U.S.	Communication development, new technologies and social changes in information communication, social movement and communication, communication and sustainable development, and related issues	Modernization theory (innovation diffusion), criticism theory, power communication or free communication theory (theory of participation in communication, theory of entertainment education)
Health communication	1970s	U.S.	Social marketing and health promotion campaign related to health improvement; communication between doctors and patients and medical technology promotion related to disease diagnosis and treatment; risk health information communication related to risk communication	Two-level communication, innovation diffusion, persuasion communication, social marketing, social learning

(continued)

Table 1 (continued)

Communication mode	Time of origin	Place of origin	Research content	Main theory
Science communication	1930s	U.K	Communication process and mechanism of scientific knowledge	Communication-related theories
Risk communication	1980s	U.S.	Risk communication based on theoretical exploration and practice specification, subjects based on the communicator and audience, communication mechanism, public awareness	Theories of risk society, psychology and management
Political communication	1920s	U.S.	Political information of political activity, news media, public opinion, new media, etc.	Political rhetoric theory, agenda setting theory, Spiral of Silence, persuasion theory

Source Prepared by the author

formed yet. Currently, we should focus on accumulating empirical research experience and trying interdisciplinary researches, instead of constructing theoretical systems and research frameworks hastily. This point also applies to climate change communication.

Third, the above-mentioned communication modes all encounter the interdisciplinary dilemma. Take political communication as an example. Political communication can be defined from the perspective of politics or communication, but horizon fusion or introduction of interdisciplinary research methods is insufficient. Climate change communication involves politics, economy, society, environment, development and other fields. For in-depth researches, we should strengthen the introduction, reference and verification of interdisciplinary research methods on the basis of paying attention to the horizon fusion between climate science and communication science.

Finally, climate change communication and the above-mentioned communication modes all originated from Europe and the U.S., especially the U.S. From the perspective of constructivism, the "national conditions" such as political, economic and social factors would be inevitably taken into account in the research and development process. While learning and absorbing, scholars from different countries need to make the researches really localized according to their own national conditions.

2.3 International Research on Climate Change Communication

When climate change was discussed only as the environmental issue, Western media had started explorations on climate change communication. In the 1970s, the term "global warming" began to appear frequently in the media report. On July 21, 1977, full-time writer Paul Valentin wrote an article entitled "Hotter and hotter in the next 100 years" for the *Washington Post*. On February 18, 1978, Thomas Thule published an article entitled "Climate experts predict global warming trend", in which he, for the first time, said that consumption of coal and oil leads to the concentration increase of carbon dioxide in the atmosphere, which will make global temperature rise. Thomas Thule can be regarded as the first person to accurately describe global warming and climate change.

In the 1980s, climate change became a global issue after the continuous research by European and U.S. scientists. A debate on the authenticity of climate change was sparked, lasting for more than a decade. After a survey, Moser found that those skeptics about climate change were mainly the representatives of traditional fossil energy enterprises, who, to safeguard their immediate interests and continue to seek exorbitant profits, bribed some scientists and think tanks and spread the misleading news (climate change is false, exaggerated, without any consensus in science) through the designated media, in an attempt to reverse the public awareness of climate change.

Of course, some people firmly believed in the existence of climate change, who refuted the skeptics with strong evidences obtained after long-term data collection, emphasizing the necessity to address climate change. They voiced their opinions through the media, unconsciously acting as climate change communicators, and making more people know about the latest climate change, principle and other scientific knowledge.

This tit-for-tat debate lasted for a long time. Undoubtedly, the media was unwilling to miss such a wonderful contest and made comprehensive reporting. However, debate topics were all complex scientific issues, and the media focused on reporting the academic controversy and could only repeat the conclusions of both sides in the face of very professional fundamental research on climate change, and complex data analysis. As a result, public awareness of climate change varied a lot when the media expressed different views on climate change. In this period, the media only made superficial reporting of climate change due to its scientificity and speciality, thus failing to help the public fully aware of the climate urgency.

After more than 20 years of in-depth research on climate change, scientists have reached more and more consensus on the causes, impacts and countermeasures for climate change. The research on climate change, especially on the impacts of climate change, has made the complex scientific issue closer to the general public. After experiencing the extreme weather and climate events such as extreme cold, extreme warmth, drought and flood, the public has profoundly realized the impacts of climate change. As the understanding of climate change is deepened, the public has begun to actively think about how to deal with climate change. Climate change communication is no longer a "contest between experts". Although the debate on climate change is still going on, the public gives more attention to countermeasures.

According to a public opinion survey made in 1988, "people in the U.S., Europe and Japan are increasingly worried about climate change" (Leiserowitz, 2007/2008). Public awareness of climate change directly influences the policy making. Studies about public awareness of climate change support scientific decision making. Some researchers have found communication rules hidden behind public awareness data. Other researchers focus on analysis of content, text and discourse framework.

Along with the deeper understanding of climate change issue, more and more attention has been given to climate change communication as never before. In 2004, the U.K. government entrusted the relevant department to conduct a national climate change communication strategy survey. In 2005, the U.K. government planned to allocate GBP12 million for the national climate change communication program, with an aim to arouse the public attention to climate change, so as to address climate change through universal emissions reduction. In February 2006, the strategy report entitled "Tomorrow's Climate, Today's Challenge" was released, in which the establishment of a special climate change challenge fund was officially proposed to support climate change communication at the national and regional levels. At the same time, some environmental organizations also initiated the "national climate change campaign" for the same purpose. Subsequently, climate change communication became a cutting-edge interdisciplinary subject. The U.K., Germany, Sweden, the U.S., Canada and other countries that started climate change researches early set up special funds to support climate change communication researches. As well, some universities and research institutions established climate change communication research centers. Research results of various countries show that the current researches on climate change communication lay particular stress on the practical methodologies and audience psychology, to enhance the public awareness of climate change. For example, in the climate change communication strategy report, the EUSPACE Program proposed the practical standard for climate change communication, i.e., breaking the myth of climate change, thinking in an innovative way, effectively linking politics and communication, following the audience orientation principle, and conducting type analysis and effective management. With the support of the National Science Foundation of the U.S., Colombia University completed the masterpiece *Climate Change Communication Psychology*, which elaborated eight principles of climate change communication: fully understand audience demands; attract the audience's attention; convert obscure scientific data into concrete examples; avoid excessive use of emotional appeals, emphasize the uncertainty of science and climate change, fully develop the associations between individuals, encourage group participation and reduce the action difficulty. Thanks to such researches, climate change issue was spread rapidly all over the world and became a global concern at the end of 2009 (see Fig. 1).

In short, the researches of EU and U.S. on climate change communication mainly focus on public perception, media reporting discourse framework and communication strategies.

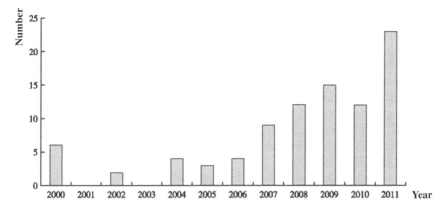

Fig. 1 English literature of climate change communication published in the academic search premier and Scopus databases from 2000 to 2011 (Wibeck, 2014: 387–411). *Source* See English references [94] for details

1. Public perception research

Before 2000, most scholars interested in public perception focused on public response to environmental issues. After 2000, more and more scholars began to pay attention to public climate change cognition and better climate change communication from the perspective of social sciences. Relevant research topics include the public awareness of climate science (Etkin & Ho, 2007), public attitudes toward different behavioral strategies for climate change (Ohe & Ikeda, 2005) and obstacles to public participation in climate change (Lorenzoni et al., 2007), etc. These study results have been applied in different climate change communication campaigns, such as the EU's Climate Action Campaign since 2010 and the one-year Swedish Climate Campaign initiated by Swedish Environmental Protection Agency in 2002, etc.

Studies show that although the climate change awareness of people in many countries is much enhanced than that 20 years ago (Corbett & Durfee, 2004; Whitmarsh, 2011), the people in some countries paid less attention to climate change around 2011 (Leiserowitz et al., 2011a; Poortinga et al. 2011; Whitmarsh, 2011). According to a climate change cognition survey conducted by YPCCC in 2011, the proportion of people "very worried" and "somewhat worried" about global warming fell from 63% in 2008 to 52% in 2011. Studies show that some Americans have "aesthetic fatigue" on climate issues (Maibach et al., 2010). In contrast, the situation in the U.K. is different. According to a questionnaire survey in 2012, about 71% of U.K. respondents are very or relatively concerned about climate change.

Lorenzoni and Pidgeon studied the climate change cognition of the European and U.S. people from 1991 to 2006. The results show that the rise or fall of the public awareness ratio is cyclical. The respondents were fully aware of the existence of climate change but lacking the cognition of causes and solutions. Climate change is deemed as a serious threat but is still far away in time and space. Therefore, they believe that climate change is less important than other personal or social risks, thinking that the governments, rather than individuals, should be mainly responsible. Lorenzoni and Pidgeon concluded that early researches prove that the general public is contradictory about climate change. Wolf and Moser (2011) also made similar conclusions through literature review.

In summary, although climate change has attracted public attention, its complexity and uncertainty still make the public awareness ambiguous (Campbell, 2011; Donner, 2011; Featherstone et al., 2009). The governments and the international community expect the public participation in climate change actions (Ockwell, Whitmarsh, & O'Neill, 2009), so climate change communication aimed to promote public participation is very important.

By area, most public awareness studies are concentrated in the U.S. and the U.K., and few others scatter in Norway (Ryghaug, Sorensen, & Naess, 2011), Sweden (Olausson, 2011; Sundblad, Biel & Garling, 2008; Uggla, 2008), Malta (Akerlof et al. 2010), Canada (Akerlof et al., 2010), Japan (Ohe & Ikeda, 2005; Sampei & Aoyagi-Usui, 2009) and Australia (Bulkeley, 2000; Herriman, Atherton & Vecellio, 2011). Most studies are mainly specific to single countries. Although there are some cross-country comparisons (Akerlof et al., 2010; Lorenzoni & Pidgeon, 2006; Wolf & Moser, 2011), only few ones involve the cross-country comparisons of developing countries (Wolf & Moser, 2011). There are insufficient studies on public awareness of people in developing countries, and comparisons between developed countries and developing countries.

2. Discourse frame analysis of media reports

Mass media coverage has an important impact on public perception of climate change. A lot of literatures emphasize that media such as television, newspaper and network act as a bridge between scientists and the public, playing a decisive role in influencing the public's understanding of scientific issues (Kahlor & Rosenthal, 2009; Olausson, 2011; Ryghaug, Orensen & Naess, 2011; Zhao et al. 2011).

The contextual analysis of climate change reporting is also one of the focuses (Gelbspan, 2005; Carvalho, 2005; Becker, 2005). Becker compared media reporting in the U.S. and Germany, arguing that U.S. journalists are more interested in the political context of climate change, while German journalists pay more attention to the creation of environmental context.

In addition, there are many discussions on journalistic practice, such as the principle of balanced reporting (Boykoff, 2005; Gelbspan, 2005; Tolan & Berzon, 2005). Balanced reporting gives the same voicing opportunity to climate change authenticators and skeptics, but to a certain extent, makes the public falsely think that they have the same influence and scale. Balanced reporting is one of the reasons why climate change skepticism was popular among U.S. people and policy makers in the twentieth century (Schweitzer, Thompson, Teel & Bruyere, 2009). Boykoff found that the balanced reporting by U.S. media from 1990 to 2002 amplified the voice of climate change skepticism, while resulting in inadequate coverage of anthropogenic climate change. As a result, the public mistakenly thought that the scientific circles have not yet reached a consensus on anthropogenic climate change. It can be seen that the principle of balanced reporting to ensure the objectivity of reporting might trigger new information biases. Thus, Boykoff suggested replacing the principle of balanced reporting with the principle of evidence proportion.

Framework analysis is also the primary method for studying mass media reporting of climate change issues. The media of different countries vary significantly in climate change reporting framework. For example, the media of Sweden, France, and Germany prefer reporting of "certainty", believing that anthropogenic global warming is the main cause of climate change, and the consequences are visible (Olausson, 2009). Instead, the U.S. media emphasize "uncertainty" to reduce public attention to climate change (Nisbet, 2009). It is worth noting that with the increase of extreme weather and climate events in the U.S. in recent years, the U.S. media have seldom reported the "uncertainty" of climate change gradually (Zhao et al. 2011). In addition, the media often adopt the frameworks of economic development (emphasizing that climate change is a new opportunity to drive economic development), advantages and disadvantages (emphasizing the positive or negative effects of climate change, negating or underestimating other impacts), ethics (emphasizing respect for nature), and public health, etc. (Adam, 2012; Nisbet, 2009).

It is worth noting that the mass media studied by European and U.S. scholars are not only limited to news media, but also include comedy, historical programs, meteorological programs, talk shows, documentaries, and children's programs, etc. For example, some scholars studying mass culture emphasize the significant role of Hollywood film "The Day After Tomorrow" in public awareness of climate change (Balmford et al. 2004; Leiserowitz, 2004; Lowe, Brown, & Dessai, 2006).

In addition, relevant literatures show that in the new media era, most of the studies are still focused on traditional mass media, rather than social new media (Koteyko et al. 2010).

3. Study of climate change communication strategies

The study of international climate change communication strategies can be roughly divided into two categories, namely macroscopic analysis from different perspectives such as philosophy and culture, and direct offer of targeted microcosmic strategy suggestions.

Take the first category for example. Moyer (2005) thinks that relevant words should be selected for climate change communication after understanding the beliefs of different audiences, and unfamiliar scientific terminology should not be used directly. For example, if the audiences are Christian, metaphorical words with spiritual color should be selected for better communication effect.

Russill (2007) distinguishes four types of rhetorical strategies of climate warning, pointing out that the climate change communication should be based on the positive and constructive consensus. Von Storch and Krauss (2005) compared the responses of U.S. and German audiences to meteorological disasters, emphasizing the importance of internal cultural factors to climate change communication.

Moser and Dilling (2004) put forward seven microscopic strategies for climate change communication, such as subdividing audiences, selecting appropriate information to amplify the credibility and legitimacy of climate change, selecting appropriate communication channels, linking climate change with people's lives, and stimulating actions on emissions reductions through cultural values and beliefs. Studies show that U.S. people are more sensitive to the information of competitiveness, leadership, originality, innovation, justice and public well-being. Freimond (2007) specially designed a practical climate change communication strategy for the companies to prevent them from being mired in the scandal of "green washing".

4. Summary

As shown above, fruitful results have been achieved in international climate change communication researches, especially in the past 10 years, but there are also obvious deficiencies.

First of all, international climate change communication researches focus on communication and behavioral psychology. Existing researches are usually based on the theoretical basis of communication and psychology, focusing on applied research, and lacking broader academic perspectives, such as international relations and stakeholder analysis, etc.

Second, public awareness research is a fast-growing area of international climate change communication research now, but it still stays at the level of audience segmentation. Although certain practical explorations are made in promoting local actions, there are few follow-up studies to stimulate more actions.

Finally, current researches on media content, public awareness or strategy, are mostly concentrated at the developed countries, giving little attention to developing countries, especially vulnerable groups in developing countries.

2.4 China's Researches on Climate Change Communication

In order to more comprehensively sort out the literature on China's climate change communication research, I found out a total of 81 articles published from 2007 to 2014 after entering the keywords of "climate", "report", "news", "framework", "cognition" and "strategy" in the full-text databases of China National Knowledge Infrastructure (CNKI) and Wanfang. By content, they are divided into four categories: overview (6 articles), media content and discourse framework (49 articles), communication subject role and strategy analysis (7 articles), and public awareness (21 articles). These literatures outline the basic situation of climate change communication research in China (see Fig. 2).

From 2015 to 2017, the literatures on media report analysis were greatly improved in quantity and quality and maintained at around 30 pieces per year, higher than in 2010, when there were 20, but there were few studies on public awareness and communication strategies.

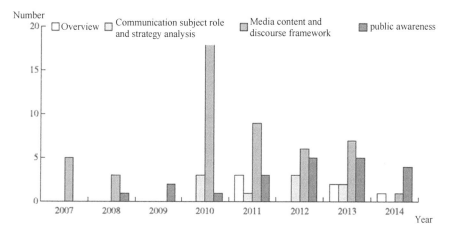

Fig. 2 Chinese literature on climate change communication collected in full-text databases of CNKI and Wanfang from 2007 to 2014. *Source* Prepared by the author

1. Research on media reporting discourse framework

Climate change had become a global issue in the 1980s, but China's media and academic circles have not noted it until more than 20 years later. By time sequence and attention degree, it can be divided into four stages.

The first stage lasted from 2007 to 2009. As a series of international hot events occurred, including the celebration of the tenth anniversary of the signing of the Kyoto Protocol, the issue of the fourth assessment report by the IPCC, and the launch of the Bali Road Map, the international community showed more attention to climate change in 2007. As a key player in global climate governance, the Chinese government participated in the COP13 in Bali, which was witnessed by few Chinese media that reported the international community's attention to climate change. At the same time, China's National Assessment Report on Climate Change and China's National Climate Change Programme were issued, expressing China's determination to address climate change, and attracting further attention to it.

In June 2007, Hepeng Jia published an article entitled "Global Warming, Science Communication and Public Participation—Science Communication of Climate Change in China" analyzing climate change reporting by *People's Daily, Science and Technology Daily, Science Times and Beijing News* in 2005 and 2007. In August 2007, *Chinese Journalist* published feature articles titled "Climate Change and Media Responsibility", which were written by the journalists with experience in climate change reporting, putting forward specific suggestions on the ways and means of climate change reporting (Chen, 2007; Liu, 2007; Feng, 2007; Xu, 2007). In 2008, several papers were also published, but their topics were scattered (Jiang, 2008; Ren, 2008; Jia, 2008).

The literatures at this stage have three characteristics: authors are journalists engaged in climate change reporting; the papers are mostly based on practices, aiming to popularize the climate change knowledge and reflect on journalists' attainment; the orientation of climate change reporting is quite chaotic. For example, science journalists focus on science communication, and environmental journalists highlight environmental communication, all without independent research orientation awareness.

The second stage lasted from 2010 to 2011. In December 2009, the COP15 in Copenhagen became a focus event because of its importance of issues, large number of participants, and high level of attention. Nearly one hundred reporters from China witnessed this historic event at the Bella Center, conducting interviews from different angles, and making a large number of on-site reports. Wei Su, deputy head of the Chinese delegation and director of the Climate Change Department of the National Development and Reform Commission, once reviewed this event: "The main achievement of COP 15 is that people's cognition of climate change has been enhanced. In just two weeks, with the help of some modern media technologies, Copenhagen became the focus of the whole world." (Su, 2009). Some commentator said: "The COP15 in Copenhagen achieved a global agenda setting for climate change." (Yu, 2010). Since 2010, a number of academic achievements have been published, which focus on the media coverage and discourse framework of Copenhagen Climate Change Conference.

Situation at this stage is significantly different from the first one. Besides the news analysis articles written by the journalists, the scholar in the field occupied a larger proportion, which deepened the theoretical discussions about the COP15. For example, Xiaoping Guo and other scholars (Guo 2010; Zhang, 2010; Qian, 2010) paid attention to international media's reporting of China, critically analyzing the discourse framework of international media and revealing that some of them have prior hypothesis bias against China, which affected the agenda setting. Another group of scholars (Jiang 2010; Qu, 2010; Jia, 2011) conducted a comparative study of domestic and international media's reporting on COP15 and found the ideological gaps and national condition differences between them.

The third stage ranged from 2012 to the first half of 2015. In 2012, the enthusiasm of discussions on COP15 was weakened and returned to normal, and there were no much theoretical breakthroughs compared to the second period. Relatively speaking, scholars at this stage were more aware of climate change, consciously including their researches into the scope of "climate change communication" and trying to analyze problems from a macroscopic perspective. For example, Liu Tao clarified the research orientation in the article of "Theoretical Research on Rhetoric of New Social Campaign and Climate Change Communication" citing Prof. Baowei Zheng's definition of climate change communication, introducing the concept of risk society, thinking the climate change topic characterized by uncertainty has penetrated all uncertain social fields, and pointing out that climate change communication is a typical new social campaign to "deeply inject the public values of fairness, equality, justice, tolerance, dialogue, etc. into the minds of people." (Liu, 2013).

The fourth stage started after the COP21 in Paris in 2015, which once again attracted the attention of global media. A large number of reports were published, becoming new materials for scholars' research.

From 2015 to 2017, analyses of media coverage continued to keep at a steady level.

2. Research on the communication subject role and strategy

Compared with the research on media reporting, the researches on the communication subject role and strategy are limited. The researchers are mainly several scholars from the China Center for Climate Change Communication (China4C). However, the researches on the communication subject role and strategy have made special contributions to the climate change communication research in China.

Based on the analysis of communication subject role and corresponding strategy, Chinese researchers clearly point out the importance of climate change communication research. At the COP15 in 2009, Chinese government, media and NGOs made their group debut. Compared with the developed countries with rich experience in the international cooperation among different stakeholders, China has a long road to improve its win–win cooperation capacity.

Chinese scholars, in particular Baowei Zheng and myself, pay attention to practical problems and solutions, participate in and follow up the subsequent UN Climate Change Conferences, conducting tracking analysis of role and strategy of climate change communication subject (Zheng &Wang, 2010, 2011, 2013). In 2011, Prof. Zheng wrote the book *The Theory and Practice of Climate Change Communication*, in which he formally proposed the concept and theoretical framework of "climate change communication".

As a major carbon emitter and a developing country, China faces the pressures of both emissions reduction and development. There is no ready-made international experience for China to solve the problems China encounters in international climate negotiations. Chinese scholars need to make objective analysis according to national conditions and international situation, so as to solve practical problems.

Chinese researches on climate change communication can be traced back to 2007, but previous studies only focused on news business and discourse framework. The research perspectives of climate change communication subject role are extended to international relations and public management, and useful explorations are made in interdisciplinary study of climate change communication.

3. Public awareness research

Chinese scholars have ever conducted public awareness surveys respectively on farmers (Yun, 2009; Lu, 2010; Tan, 2011; Xiao, 2013), urban residents (Cui, 2014), corporate managers (Xu, 2011), university students (Chen, 2008; Luo, 2009; Wang, 2012; Chen, 2012) and the public in different areas (Chang, 2012; Cao, 2014), and further put forward the corresponding policy and action suggestions.

In 2012, China4C conducted a public survey on 4,169 respondents from urban and rural areas of China (excluding Hong Kong, Macao and Taiwan), to comprehensively understand Chinese people's perception, attitude and practice about climate change and other related topics. Only 6.6% of the respondents said they never heard of climate change. Most people think that climate change is occurring mainly because of human activities, and China has been exposed to climate change hazards, which have greater impacts on rural residents (Wang, 2014). This is the first national public survey conducted by an independent third party, which provides a data reference for international negotiations and the domestic policy making.

Based on the surveys, I compared the data of the U.S., Mexico, China, Taiwan, and seven Asian countries, and found some common points. For instance, the general public's cognition of climate change remains at the subjective assumption level based on personal experiences, with dominating emotional components; for most people, there is still a gap between climate change cognition and action; although new media are booming, television is still the main channel for climate change communication (Wang, 2014). Other scholars from China4C also conducted in-depth analysis from different angles (Li, 2013a, 2013b).

In 2013, China4C conducted a low-carbon awareness survey for urban residents, and summarized four types of low-carbon cognitions, behaviors, their basic characteristics according to the survey results and four indicators (low-carbon concept cognition, low-carbon policy cognition, low-carbon willingness to pay and low-carbon behavioral expression), proposing the targeted media communication strategy in consideration of "four types of low-carbon people" and their characteristics (Zheng &Wang, 2013a).

After the first nationwide public climate cognition survey in 2012, I hosted the second one in 2017. The computer-aided telephone survey covered 4,025 respondents from 332 prefecture-level administrative units and 4 municipalities in Mainland China, paying special attention to urban–rural ratio and gender ratio to more objectively reflect the general public awareness situation. Survey results show that Chinese people's climate awareness remains high (94.4% of respondents believe that climate change is happening, 66% of respondents think that climate change is mainly caused by human activities, and 79.8% of respondents are worried about climate change). Chinese people highly appreciate the relevant government policies to mitigate and adapt to climate change. Compared with five years ago, air pollution and health problem have become the climate change impacts that Chinese people are most worried about. Technological innovations such as bicycle sharing have provided smart solutions for public participation in response to climate change. Based on the survey results, the research team is conducting in-depth studies (Wang, 2017, 2019, 2020).

4. Summary

Chinese literature reviews show that most of climate change communication researches still focus on media content analysis, which is greatly related to the research content availability and the researcher's knowledge system. The UN Climate Change Conferences in 2009 and 2015 attracted a lot of media. Subsequently, the studies on media reporting contents showed a blowout trend. Both climate change reporting and climate change communication research have topical characteristics, lacking continuous attention within a certain period.

Table 2 Comparison of climate change communication researches in China, Europe and the U.S.

Country Climate change communication research	Europe and the U.S.	China
Start time	1980s	Early twenty-first century
Research origin	Debate on climate change authenticity	COP15 in 2009
Basic attitude	Climate change was regarded as an environmental issue; paid attention to emissions reduction	Climate change was regarded as a development issue; paid balanced attention to emissions reduction and adaptation
Research orientation	Environmental communication	Climate change communication
Research framework	Media content and discourse framework analysis, public awareness, strategy analysis	Media content and discourse framework analysis, analysis of communication subject role and strategy, public awareness
Research methods	Quantitative analysis supplemented by qualitative analysis	Qualitative analysis supplemented by quantitative analysis
Research representative	YPCCC	China4C

Source Prepared by the author

Comparison results show that compared with the climate change communication research in Europe and the U.S., climate change communication research in China started late and is significantly different despite similarity in the research framework. China's climate change communication research is an initiative research based on existing experience and national conditions. Some scholars point out: communication of climate change knowledge and information is the research subject of climate change communication in China and the West, but "they are different in start time, research origin, basic stance towards climate change, research orientation, framework and methods." (Zheng &Wang, 2013) (see Table 2 for details).

Chapter 3
Two-Level Games of China

According to the two-level game theory, the state head of a country needs to address both national and international challenges and balance the interests of different interest groups to obtain the maximum win-set. In addition, the two-level game emphasizes the interaction between international and national factors, rather than simple superposition of them.

In recent years, China has played an increasingly important and crucial role in global climate governance, with its global influence enhanced year by year. China's attitude to climate change is related to China's overall economic and social development, as well as the success or failure of global climate governance. Only by taking into account both international and national factors can China better play its due role and promote the global climate governance. As a strategic tool for climate governance, climate change communication research should be based on the two-level analysis framework and two-level game theory, to more accurately transmit information.

This chapter will mainly analyze the two-level game motivation and objects of China in response to climate change and clarify the analytical framework for the following chapters.

1 Motivation

The two-level game theory emphasizes that the international and national factors are not isolated but interactive. The main participants of the UN climate negotiations are state members. The state is the basic legal subject of international climate legislation. The main principles, norms, rules and decision-making procedures of international climate governance are established mainly through negotiation between state members (Ge, 2005). The attitudes and policies of states play a crucial role in the effectiveness and authority of international climate legislation. All states should implement the policies of international climate governance to ensure the normal operation of the international climate governance mechanism. So far, the state is still the most important actor of global climate governance.

According to the two-level game theory, the source of international issues can be found at the domestic level. China's actual needs for development drive its active engagement in global climate governance. At the same time, the motivation of domestic issues can be found at the international level. China attaches great importance to climate change response, which is attributed to the pressure of the localization of institutional climate rules and regulations.

1.1 Domestic Roots of International Problems

First, China is a direct victim of climate change.

The state seeks its maximized interests through various ways in the international game process. A country affected by climate change seriously is more willing to participate in global climate governance. According to the State Meteorological Administration, China is much affected by the monsoon, so the climate in China is complex and variable. In addition, different regions of China have different climates. Since the mid-twentieth century, China has been increasingly affected by climate change. For example, warming magnitude in China is twice of that of the global average, and there are more and more extreme weather and climate events such as drought and flood. On November 20, 2015, China's Ministry of Science and Technology released the Third National Assessment Report on Climate Change, which shows that since 1909, China's warming rate has been higher than the global average, and the average temperature has risen by 9–1.5 °C per 100 years. From 1980 to 2012, the sea level of China's coastal areas rose by 2.9 mm/year, higher than the global average. From the 1970s to the early twenty-first century, the area of China's glaciers and permafrost reduced by about 10.1 and 18.6%. In the future, the temperature in China will continue to rise. By the end of the twenty-first century, the average temperature of China may rise by 1.3–5 °C, the frequency and intensity of extreme events such as rainstorm, strong wind, flood, storm tide and wide-ranging drought will be enhanced, and sea level will continue to rise.

Long-term observations show that the overall impact of climate change on China's food security is unfavorable. Climate change has an impact on water resources, and available water resources fall short of demand. Climate change has also destroyed the stability of ecosystem, leading to soil erosion, ecological degradation, and species migration. Although the nationwide weather phenomenon of smog in the recent years is not directly caused by climate change, climate change that can reduce the atmospheric environmental capacity is not conducive to the diffusion of pollutants, and to a certain extent, becomes the "accomplice" of smog. The national survey conducted by China4C in 2017 shows that more than 70% of Chinese people agree that climate change and air pollution affect each other.

1 Motivation

As estimated, further warming will have a major adverse impact. China's natural disaster risk is at a global high level, and is highly sensitive to climate change, whose adverse impacts have systematically penetrated the economy and society.

As a direct victim of climate change, China actively participates in global climate governance, which can draw on the international advanced experience in addressing climate change.

Second, China needs to strive for development space.

In the process of formulating the UNFCCC and subsequent international regulations and agreements, the participating countries faced the core contradiction between their own economic and social development and global response to climate change.

In the production and consumption process, traditional fossil fuel emits a large amount of greenhouse gas, which directly leads to global warming. However, it is undeniable that production and consumption of traditional fossil fuel greatly promoted the economic and social development. China undergoing new urbanization faces many problems such as industrialization, transfer emission and energy structure adjustment, etc. China is carrying out large-scale infrastructure construction in the areas of construction and transportation, etc. and low efficiency of technology leads to the low efficiency of energy consumption. In a certain period of time, coal-based fossil fuel will be still China's main energy source, and China will remain as the world factory. Developed countries have transferred large amounts of greenhouse gas emissions to China by setting up heavily polluting factories in China.

Although having become the second largest economy in the world, China still faces the problems of insufficient and imbalanced development in the new stage of socialism. The problem of urban–rural differences in China is prominent. In the process of rapid urbanization, per capita energy consumption of urban residents has been nearly three times as much as that of rural residents, which inevitably leads to a rapid growth of energy consumption. According to the latest statistics of the National Bureau of Statistics, at the end of 2017, China had a rural poor population of 30.46 million. As the most direct victims of climate change, poor people emit the least greenhouse gases, but are most vulnerable to extreme weather and climate disasters. Studies show that in China, ecologically fragile areas, poverty-stricken areas and climate-vulnerable areas are highly superposed (Oxfam, Greenpeace, Chinese Academy of Agricultural Sciences, 2009. See Fig. 1 for details). Climate change has made it more difficult for China to carry out environmental protection and poverty reduction.

Precisely because of these realistic problems, China needs to actively participate in international climate negotiations and ensure its own development in the process of international system construction and evolution.

As well, China should strive for the right to formulate global climate governance rules.

China's participation in the global climate governance is different from involvement in other international systems. Compared to global governance in the economy and other fields, global climate governance is a late comer, whose mechanism is still under construction and needs to be improved by the international community.

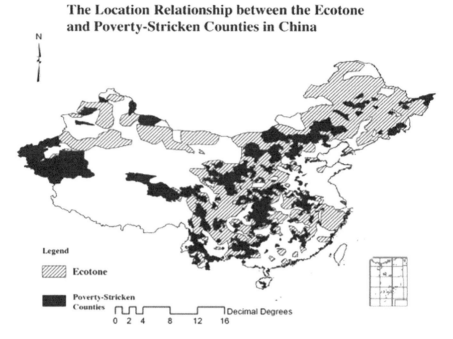

Fig. 1 In China, ecologically fragile areas, poverty-stricken areas and climate-vulnerable areas are highly superposed (Oxfam, Greenpeace, Chinese Academy of Agricultural Sciences, 2009), drafted by Yinlong Xu

China should strive for the rulemaking right of global climate governance, to ensure more equal participation of developing countries in global climate governance. At the same time, accumulated experience in climate governance will be of reference significance to governance in other areas.

1.2 International Roots of Domestic Problems

International climate regulations stipulate the objectives and obligations of contracting states in addressing climate change, clarifying the specific requirements for developed countries in emissions reduction. Contracting states are required to take intervening measures at home to meet international requirements. However, they will encounter many challenges in the localization process of international climate regulations and need to adjust the internal structure to ultimately localize international regulations.

First, the combined action of institutional pressure and interest perception is a prerequisite for the localization of international regulations. China faces big institutional pressures at both the international and domestic levels. On the one hand, China faces the internal pressure of greenhouse gas emissions reduction and economic development. On the other hand, the pattern of global climate governance is restructured, which also exerts pressure on China. In addition to stress, impetus is also available for the international climate incentive mechanism. The clean development mechanism (CDM), carbon trading mechanism and joint implementation mechanism stimulate the relevant countries at both material and social levels. These incentive mechanisms also arouse China's interest perception of international climate systems. China has realized that participation in international climate governance is not only conducive to the improvement of its international image, but also can bring about tangible benefits through the related mechanism, i.e. CDM. In addition, China's unique domestic structure is a necessary guarantee for the localization of international regulations. China is government-driven, so that an international regulation supported by the central government can be implemented quickly. After the COP15 in Copenhagen in 2009, the Chinese government felt unprecedented international pressure, becoming more aware of the importance to take the initiative to address climate change. On this basis, the Chinese government quickly localized the international climate mechanism in China, paying increasing attention to climate change, and playing an increasingly important role in global climate governance.

2 Objects

2.1 International Game Objects

Climate change is a common challenge facing by all humans. Effective addressing of climate change is in line with the interests of all countries and humans, which requires close cooperation and joint efforts of all countries. No country can deal with global climate change alone. Therefore, establishment of an efficient international climate governance mechanism is an inevitable choice for addressing global climate challenges.

The international mechanism is "a set of implicit or explicit principles, norms, rules, and decision-making procedures around which actors' expectations converge in a given area of international relations" (Krasner, 1983). The UNFCCC and Kyoto Protocol have a milestone significance to global climate governance. The international climate governance mechanism has been initially established. Since then, the Conferences of the Parties ("COP" for short) have focused on improving the international climate governance mechanism.

According to the two-level game theory, to maintain its ruling status, the government of a country will strive for maximized profits and minimized losses at the international occasions. In the international climate negotiations, China faces the pressures from both developed countries and developing countries.

The divergence between China and developed countries mainly lies in the historical responsibility of emissions reduction. The UNFCCC stipulates that all contracting parties shall bear "common but differentiated responsibilities", and actively and jointly address the climate change to protect the Earth and safeguard the common interests of all humans and descendants. Always observing the UNFCCC, China holds that developed countries should bear the primary responsibility for climate change because they had emitted a lot of greenhouse gas during the industrial revolution, while China as a developing country and a victim of climate change, should not assume any obligation to reduce emissions. However, developed countries think that China has already been the world's largest emitter, and should address climate change and undertake binding emissions reduction obligations more actively.

The relationship between China and developing countries has undergone subtle changes with the deepening of negotiations. China needs to actively unite the developing countries and deal with the relations between developing countries. Climate change is directly related to the survival of the Alliance of Small Island States and the least developed countries, so they require the international community to reduce emissions most urgently. As a developing country, China has been the world's largest emitter. For the sake of own safety, different countries affected by climate change to different extent will show obvious divergence of attitude.

In addition to the negotiation pressure from other countries having different opinions, China has to face the pressure from international NGOs which were previously neglected. International NGOs play an independent third-party role in regulating international climate negotiations, which advocate protection of fragile states and environments, and maintaining of climate justice. To some extent, they influence climate negotiations. In recent years, China has rapidly accumulated experience in working with international NGOs in climate negotiations.

2.2 Domestic Game Objects

The two-level game theory emphasizes the game at the domestic level, holding that various interest groups force policy makers to adopt policies that are beneficial to them. For example, the U.S. are dominated by their own political parties, parliament members, institutional spokespersons, and interest group representatives, etc. In the past 40 years of reform and opening up, different interest groups have also emerged in China. A country should be aware of the bottom line of domestic game objects to more accurately predict the trend of international negotiations.

First, different central government departments have the conflict of interests, and cross-departmental coordination capabilities still need to be improved. Climate change work was first in the charge of China Meteorological Administration, which could hardly mobilize other departments, while different ministries and commissions gave different attention to climate change work. As China attached importance to climate change issues, the climate change response work was transferred to the National Development and Reform Commission which has the power to make decisions on energy policies. In 2007, the climate change issue attracted the attention of top leaders. As a result, a leading group composed of Prime Minister and members of

18 ministries and commissions, was set up in the headquarters of National Development and Reform Commission to coordinate the work about climate change response. All ministries and commissions paid more attention to climate change, but they still encountered some contradictions because of conflicts of interests. On March 13, 2018, the State Council's institutional reform plan was submitted to the First Session of the Tenth National People's Congress for deliberation. The National Development and Reform Commission's original responsibility for climate change and emissions reduction was transferred to the newly established Ministry of Ecology and Environment. Of course, the performance of this new ministry remains to be tested, but the original cross-department coordination problem is somewhat relieved.

Second is the game between the central government and the local government. The central government has unveiled a series of laws and regulations to enhance emissions reduction. However, implementation of them will encounter resistances from the local governments.

Finally, there is a game between the government and the enterprise. Enterprises, especially high-emission enterprises have the biggest potential in slowing down climate change, but it seemingly runs counter to their pursuit of maximized benefits quickly in a short term. When making policies, the government should fully consider the initiative of enterprises to achieve a win–win situation.

2.3 Adaptative Revision of Win-Set at the Domestic Level

The two-level game theory emphasizes the win-set at the domestic level, which refers to the set of international climate agreements supported by the voters of some countries such as the U.S. China's climate change decision-making mechanism is very different from that of the U.S., and China's highest decisions about climate change are made by a cross-department coordination group. Being determined to reduce emissions, the central government adopting the top-down decision-making mechanism continuously improves the incentive and compensation mechanisms, striving to include the domestic game opponents into the largest win-set of climate change response.

Chapter 4
Analysis of Stakeholders (2009–2015)

1 Definition and Classification of Stakeholders

1.1 Definition

The two-level game emphasizes the interaction between international and domestic factors. Before study of climate change communication and governance, basic analysis unit of each level, namely the key stakeholder, should be identified.

One of the earliest definition of stakeholders comes from the Stanford Research Institute, which believes that stakeholders are related to the survival of an organization. Edward Freeman places more emphasis on the impact of stakeholders on the goal realization of an organization. In 1996, the World Bank defined the "stakeholder" of public governance as "an individual or a group that is positively or negatively affected by the outcome or may influence the outcome through intervention". For example, the stakeholders of the World Bank include borrowers (i.e. countries, regions and local governments, etc.), individuals or groups directly affected (i.e. people or communities in poor and underdeveloped areas), individuals or groups indirectly affected (i.e. NGOs, private sectors), and the World Bank's management, employees, and shareholders. The stakeholders of climate change refer to the individuals or groups that are affected by climate change and have the will, responsibility or ability to deal with climate change (Keskitalo, 2004).

Based on the above definitions, I hereby summarize the stakeholders of climate change communication as the individuals or groups that are affected by climate change and influence the process and effect of governance and communication, including national policy makers, scientific researchers, international and non-governmental organizations, enterprises, media, practitioners in relevant industries, and the general public (especially the poor people in ecologically fragile areas), participating in climate change communication and governance at both the international and domestic levels.

1.2 Classification Method

Although the solution of problems is inseparable from the support of all stakeholders, stakeholders are different in contribution. Stakeholders might have "very different requirements and objectives" (Deng, 2011). Scholars classify the stakeholders according to the multidimensional classification method and Mitchell score-based approach, which are most influential.

The multidimensional classification method emphasizes the character differences of stakeholders in different dimensions. Some scholars study the differences of stakeholders from the dimensions of ownership, economic dependence and social interests. Some people divide stakeholders into direct stakeholders and indirect stakeholders. By contractual relationship with the enterprise, the stakeholders may be divided into contractual stakeholders and public stakeholders.

In the late 1990s, Mitchell from U.S. proposed a scoring method based on legitimacy, power and urgency. The score decides whether an individual or a group is a stakeholder, and also its category. Legitimacy means whether a group is endowed with the legal, moral or specific claim to the enterprise; power means whether a group has the status, ability and corresponding means to influence the decision-making; urgency means whether the demands of a group can immediately attract the attention of the management. If one or more of the above three requirements are met, the target can be identified as a stakeholder of an enterprise. By the score obtained after assessment of the above three indicators, stakeholders can be classified into definitive stakeholders, expected stakeholders, and latent stakeholders. The three categories of stakeholders are dynamically converted to each other at any time with the changes in indicators. The scoring method makes the classification of stakeholders more operable and facilitates the practical classification, thus becoming the most popular method for classification of stakeholders.

This book analyzes the stakeholders of climate change communication from 2009 to 2015 in combination with the scoring method, supplements and revises the scoring method according to the characteristics of China's climate change communication at the international and domestic levels, adding a new indicator—correlation, which makes the scoring method more adaptable in the China's context.

Correlation refers to the relationship between the stakeholder and the climate change communication target and is divided into positive correlation and negative correlation. Positive stakeholders share the basic value goals and can form the joint forces through adjustment and adaptation; negative stakeholders have potential divergence in value goals and may become the resistance. The adjustment and adaptation involve the interests of other stakeholders, and the correlation is also a key indicator that reflects the interest relations of different stakeholders.

The ultimate goal of climate change communication is to address climate change issues. Climate change is an unprecedented challenge facing mankind. To solve the problem of climate change, it is necessary to identify the stakeholders who are highly like-minded and complementary in resource advantages first and amplify the positive

1 Definition and Classification of Stakeholders 53

correlation through the adjustment and adaptation, so as to reduce the game cost at two levels and obtain the largest win-set.

2 International Stakeholders

The annual UNFCCC COP is the most important stage of climate change communication. The UNFCCC as the overall framework for international climate negotiations was adopted at the UN Conference on Environment and Development in Rio de Janeiro, Brazil in 1992, which is the first international convention to address global change. Since then, the UN COP on climate has been held for more than 20 years.

Many countries and organizations have joined in the international climate negotiations, which cover the topics of mitigation, adaptation, capital, technology, finance and others of climate governance. The UN Climate Change Conference has become a professional platform for the international community to discuss global climate governance. The very important, dramatic COP15 in 2009 in Copenhagen attracted worldwide attention, making the professional climate negotiations closer to the general public. Through the constant negotiations and the spread of relevant climate change information, the UN Climate Change Conference has become the main platform for climate change communication at the international level, people around the world have deepened their understanding of climate change, and all contracting parties have engaged in the climate change communication directly or indirectly. The research in this chapter is mainly around international climate negotiations from 2009 to 2015.

2.1 Legitimacy

Legitimacy means that the stakeholder has the right to participate in climate negotiations and the willingness to spread relevant information in the process of negotiation.

According to the requirements of the UN, the UN Climate Change Conference may be attended by government representatives, independent third-party monitoring agencies and media. Among them, government representatives of different states are the main participants of climate negotiations, who formulate the main principles, rules and decision-making procedures of the international climate norms. As a sovereign state, China actively participates in international climate negotiations and legislation, and China's attitude greatly influences the effectiveness and authority of international climate legislation. China's government delegation of international climate negotiations is composed of government officials from different departments and experts from relative think tanks, so that it has the highest legitimacy in climate change communication.

Independent third-party supervision agencies refer to international agencies, universities and colleges, scientific research institutions and civil society organizations. In the frame of UNFCCC, all these above are classified as various types of NGOs. Nearly 30,000 people from NGOs participated the COP15 in Copenhagen, accounting for two thirds of the total registered participants. NGOs are divided into international NGOs and local NGOs. Among them, international NGOs have participated in the drafting and subsequent revision of important international climate change frameworks and agreements such as the UNFCCC and the Kyoto Protocol. Because of the long-term participation, international NGOs have accumulated rich experience in active negotiation, flexible participation and varied approach. Through the frequent formal or informal consultations, international NGOs supplement and improve the content of international climate agreements, playing a role in monitoring and facilitating the negotiation. From the perspective of legitimacy, international NGOs can obtain higher scores.

Since 2009, China's local NGOs have started participation in climate change negotiations. However, compared with international NGOs, they have limited capacities and limited influence on negotiation entities, only having moral legitimacy presently.

Since the COP15, the media has been a major participant of the annual COP. By interviewing government delegations, international agencies and NGOs and introducing the negotiation process in a timely manner, the media, to some extent, influences the negotiation process. Although the participation legitimacy of media is not written into international law, its identity has been recognized by the UN and all participating countries to a certain extent.

2.2 Power

Power means that a group having the information and/or channel of climate change communication has the status, ability and corresponding means to influence the decision making of negotiation entities. As negotiation entities, sovereign states undoubtedly have power.

Climate change is one of the most remarkable areas of global governance in which international NGOs are involved. On the one hand, international NGOs play the role of pressure groups, directly or indirectly exerting pressure on policy networks and groups by influencing public opinions. On the other hand, they sort and spread scientific climate change information to decision makers and the public, contributing to the reaching of consensus on climate change. Actively participating in climate negotiations, international NGOs with specialized knowledge have become an indispensable part of international climate governance, which cooperate with different actors, play a role of supervision, containment, mediation and coordination of different countries, and have soft powers different from sovereign states. Thus, international NGOs have more powers than other stakeholders except the governments. However, international NGOs lack sufficient rights to manage political and economic resources in contrast with the governments, and also lack the ability to directly influence the negotiation subjects in the face of the complex interest relations of climate governance. Compared with international NGOs, with less speciality and international activity, Chinese local NGOs are inferior to international NGOs in power because of less professional capacity on the topic and international activities.

Although the media reporting cannot directly influence the decision-making of the negotiation entities, media opinion is one of the factors to be considered by the negotiation entities when making decisions. If the media coverage involves the interests of the negotiation entities, it may immediately attract the attention of the negotiation entities, thus further influencing the decision making of the negotiation entities. Therefore, the media has a relatively strong power. However, so far, the media of China, compared to international media, is less influential and powerful in international climate negotiations.

2.3 Urgency

Urgency refers to that the claim of a group having the climate change communication information and channel is immediately concerned by the negotiation entities. As the negotiation entity, the government has the highest urgency.

NGOs will raise various claims to the government delegations at the negotiation site. If the claims are consistent with the negotiating objectives, they will immediately attract the attention of negotiation entities and accelerate the negotiation; otherwise, there will be no response. Therefore, urgency depends on the content of claim, and is dynamic.

The media is the main channel for information circulation during climate negotiations. Whether a claim can attract the attention of negotiation entities, depends on its content. In this aspect, the media and NGOs are the same.

It should be emphasized that specialized information and scientists' participation are particularly important for climate change communication. The premise of international climate negotiations is scientific prediction of climate change trends. The UN has specially established the IPCC and each participating country has its own scientific support team. Scientists are very active in international climate negotiations, but they do not play an independent role, playing their role in the capacity of members of government delegations or third-party supervisory organization delegations. Therefore, scientists are included into the scope of government delegations or third-party supervisory organization delegations in this book, rather than independent stakeholders.

2.4 Correlation

Correlation refers to the relationship between the stakeholder and the climate change communication target and is divided into positive correlation and negative correlation. Positive stakeholders share the basic value goals and can form the joint forces through adjustment and adaptation; negative stakeholders have potential divergence in value goals and may become the resistance.

The goal of China's climate change communication is constructive participation in global climate governance. As mentioned above, the government is the legal representative of a sovereign state, and the main participant and promoter of international climate negotiation. Since the official start of international climate negotiation in 1990, the Chinese government has got involved, playing an increasingly important role in the improvement of international climate change mechanism and serving as a positive stakeholder of climate negotiation. Highly consistent with the government, Chinese media is another positive stakeholder.

International NGOs are positive stakeholders in promoting the establishment of international climate governance mechanism. As the positive stakeholders of China's climate change communication, some international NGOs in China are more aware of China's practical difficulties and contributions to climate governance, playing a buffering role in front of some unrealistic expectations upon China. Of course, as the supervisors, international NGOs will also criticize the Chinese government if its governance practices are below international standards. Compared with international NGOs, Chinese local NGOs have no direct correlation with climate negotiation.

The international media's understanding of China is still relatively limited, and their reporting of China is mainly critical. Thus, correlation between China and international media needs to be enhanced.

The COP is generally composed of the main negotiation and the side events. The role of side events is to promote the exchange of knowledge and experience in climate change. Relevant governments, NGOs and media are the main players of the main negotiation, while representatives of enterprises and people mainly attend the side events. Because of not directly getting involved in negotiations, representatives of enterprises and people were relatively weak in legitimacy, power, urgency and correlation from 2009 to 2015.

To sum up, Chinese government, international NGOs carrying out climate change work in China, and international media are the definitive stakeholders of China in international level climate change communication, namely the core stakeholders. Chinese media and NGOs with the foundation of legitimacy and power are expected stakeholders, which can influence the negotiation entities more with their improvement in professional level. As latent stakeholders, enterprises and people have moral legitimacy (see Table 1 for details).

As mentioned above, international stakeholders are deemed as participants of international climate negotiation, and enterprises and people have the high legitimacy of climate change communication.

Table 1 Stakeholders of China's climate change communication and governance in international negotiations

Stakeholder type and name		Legitimacy	Power	Urgency	Correlation
Definitive stakeholder	Government (including representatives of scientists)	High	High	High	Positive
	International NGO	High	High	Medium	Positive/Negative
	International media	High	High	Medium	Positive/Negative
Expected stakeholder	Chinese media	High	Medium	Medium	Positive
	Chinese NGO	High	Medium	Medium	Low
Latent stakeholder	Enterprise	Low	Low	Low	Low
	People	Low	Low	Low	Low

Source Prepared by the author

3 Domestic Stakeholders

Climate change communication at the domestic level means communication of the latest knowledge of climate change, international negotiation pressures, relevant science and policy information to promote more active response to climate change in China.

Climate change communication at the domestic level is crucial to addressing climate change. Stakeholders involved in addressing climate change are also the stakeholders of climate change communication, including different levels of governments, scientists engaged in climate change research, international or domestic NGOs, media, enterprises, and the people affected by climate change in urban and rural areas.

According to the actual situation of domestic stakeholders, Mitchell score-based approach is revised accordingly here.

3.1 Legitimacy

Legitimacy refers to the group's specific right of participation in climate change communication and response.

The central government develops and releases national climate change policies, while different levels of local governments formulate implementation rules for national climate change policies. Legally and morally speaking, both central government and local governments have the legitimacy of participating in climate change communication and response.

Scientists engaged in climate change research (generally including scientific research workers, research groups and research institutes) are the main players of climate change research, who provide a scientific support for government policy and NGO advocacy, and an important source of information for the media coverage, having a high degree of legitimacy.

The media can be divided into central media and local media by reporting area, official media and market media by operation mode, and different kinds of professional media by reporting content. With different concerns about climate change, all types of media play a very important role in climate change communication and response and have moral legitimacy in climate change communication.

International NGOs advocate and mobilize public participation and action in climate change communication and exert pressures on policymakers and correlation networks to drive change. However, in view of the policy environment and climate change complexity of China, the international NGOs in China should not directly impose pressures, or their overall work in China will be affected. On the basis of understanding the national conditions of China, the international NGOs in China have made different adjustments to their strategies about climate change response in China to make the changes happen by cooperation, thus having moral legitimacy.

Chinese NGOs play a limited role in international negotiations and also in China due to institutional reasons. However, since 2009, Chinese NGOs have gradually owned more space in climate change response, beginning to take the initiative to play due role, which supplement international NGOs engaged in climate change response in China, and have gradually been recognized by the Chinese government, thus having certain legitimacy.

Enterprises and people of China only play a marginal role in international climate change communication but play an important role in climate change communication at home. Enterprises, especially large fossil fuel enterprises or high-emission enterprises, are the main contributors to emissions reductions, having certain participation right and legitimacy of climate change communication and response. Climate change communication and response require the public participation so the public has legal and moral legitimacy.

3.2 Power

Power means that the group has the ability of influencing policy makers' decision-making.

As the national policy maker, the central government has the greatest power in China. The power of a local government is directly proportional to its administrative level.

Chinese scientists engaged in climate change research are also major members of government's think tanks. In order to ensure the scientific decision making of National Leading Group on Climate Change, the National Expert Committee on Climate Change has been established, whose members are mainly experts in climate change science, economics, ecology, forestry, agriculture, energy, geology, transportation, architecture and international relations, etc. The National Expert Committee on Climate Change is mainly responsible for offering advices and suggestions on the scientific issues related to climate change and China's long-term strategy and major policies to address climate change. From this perspective, scientists are much more powerful than other stakeholders except the governments.

At the domestic level, the media is the main channel for climate change communication, and the government is one of the main sources of information. The media can also test the public feedback on certain policy, thus having the greater power.

The international NGOs have certain power to some extent. They engaged in climate change work in China are professional and can influence the process of policy making through collection and sorting of practice cases, submission of research reports, and policy advocacy. In contrast, Chinese NGOs have become more and more professional in the field of climate change but have relatively limited ability and means to influence policies, and relatively weak power.

3.3 Urgency

Urgency means the claim of a group having the climate change communication information and channel immediately draws the attention and response of decision makers.

At the domestic level, the central government's claim has the highest urgency. The urgency of a local government's claim is proportional to its administrative level.

Claims or suggestions of the scientists engaged in climate change research, especially those as members of government think tanks, can immediately attract the attention of policy makers, whose reasonable suggestions will be adopted immediately, with higher urgency.

The media can reflect different voices and demands from all walks of life, whose reasonable suggestions will be followed by the government, but are inferior to those of governments and scientists in urgency, although the media outshines other stakeholders.

International NGOs as active participants of global climate governance influence a country's climate change response policy and climate governance progress. However, policy makers will consider domestic situations to respond to the claims of international NGOs at the proper level.

Chinese NGOs, enterprises and general public have limited ability to directly influence policy makers because of limited capacity and channels.

3.4 Correlation

Correlation refers to the relationship between the stakeholder and the climate change communication target and is divided into positive correlation and negative correlation as same as the international level.

Climate change communication of China aims to promote the localization of international climate governance mechanism and mobilize public participation in climate change response. The two points are advocated by the central government of China, with very clear correlation. Scientists, the media and NGOs are positive stakeholders in this regard.

For the sake of local interests, local governments somewhat reduced important measures to address climate change such as emissions reduction around 2009, accordingly reserving correlation. It is worth noting that as the central government shows more determination to reduce emissions, local departments will become positive stakeholder sooner or later.

Relatively speaking, high-emission enterprises of China are the major game opponents of the Chinese government because they oppose emissions reductions most intensely, and their correlation is the weakest. However, as long as the central government is determined to reduce emissions and improves the relevant incentive mechanism, all domestic game opponents can be transformed into positive stakeholders of active cooperation.

It is observed that all domestic stakeholders have high legitimacy and the right to participate in climate change communication and response. Among them, governments, scientists and the media are the definitive stakeholders of China in climate change communication, namely core stakeholders. Relatively weak in power and urgency, international NGOs carrying out climate change work in China and Chinese NGOs are expected stakeholders. The enterprises and people only have legitimacy, and weak power and urgency, belonging to latent stakeholders (see Table 2 for details).

4 Three Major Stakeholders: Governments, Media and NGOs

The two-level game framework emphasizes the interaction between international and domestic factors. Identifying the basic analysis unit of each level is the premise of studying climate change communication and governance.

After analyzing the international and domestic stakeholders of climate change communication at present, we find that the central government of China has the highest ranking at both the international and domestic levels, showing that it as the core stakeholder is most critical and most influential in climate change response and communication in China.

Table 2 Domestic stakeholders

Stakeholder type and name			Legitimacy	Power	Urgency	Correlation
Definitive stakeholder	Government	Central government	High	High	High	Positive
		Local government	High	High to low	High to low	Positive/Negative
	Scientist		High	High	High	Positive
	Media		High	High	High	Positive
Expected stakeholder	International NGOs carrying out climate change work in China		High	Medium	Medium	Positive
	Chinese NGOs		High	Medium	Low	Positive
Latent stakeholder	People		High	Low	Low	Positive
	Enterprise		High	Low	Low	Positive/Negative

Source Prepared by the author

At the international level, the Chinese government, international media and international NGOs are the definitive stakeholders of China, ranking among the best in legitimacy, power and urgency of climate change communication; relatively weak in power and urgency, Chinese NGOs and media are expected stakeholders; the enterprises and people are latent stakeholders.

At the domestic level, the central government, local governments, scientists and media are definitive stakeholders; NGOs are expected stakeholders; enterprises and people are latent stakeholders from 2009 to 2015.

In consideration of China's objectives and stakeholders of climate change communication and governance, the central government, Chinese media and NGOs are defined as the key stakeholders of China in climate change communication and governance from 2009 to 2015 at the international and domestic levels, whose empirical research results will be presented in the next chapter through a large number of cases about their role and strategy change from 2009 to 2015 (see Table 3 for details).

Table 3 Two-level analysis of three major stakeholders

Type	Level	Legitimacy	Power	Urgency	Correlation
Central government	International	High	High	High	Positive
	Domestic	High	High	High	Positive
Chinese media	International	High	Medium	Medium	Positive
	Domestic	High	High	High	Positive
International NGOs	International	High	High	Medium	Positive/Negative
	Domestic	High	Medium	Medium	Positive

Source Prepared by the author

Chapter 5
Empirical Study: Two-Level Tracking Analysis of Three Major Stakeholders (2009–2015)

This book attempts to build a "two-level" research space to conduct a tracking analysis of three major stakeholders at the international and domestic levels. The two previous chapters discuss the research framework of two-level game and the identification of three major stakeholders. This chapter will conduct an empirical study and tracking research on the changes in the climate change communication strategies of the Chinese government, Chinese media and international NGOs at the international and domestic levels from 2009 to 2015. The core view of this book is that climate change communication is a strategic tool for climate governance, rather than just media issues. Behind the changes in communication strategy is the road of climate governance with Chinese characteristics.

I will proceed from the study of the COP15 in 2009 because this conference is a special milestone in China's history of climate change communication and governance, besides the importance of the agenda itself.

The COP13 in Bali in 2007 attracted nearly 11,000 participants, including 3,500 government officials, more than 5,800 representatives from the intergovernmental organizations and NGOs, and nearly 1,500 media journalists. The COP16 in Poznan in 2008 was attended by almost 9,300 people. The COP15 in 2009 attracted 250,000 participants, hitting a new record high in number of participants. Through various channels, the multi-stakeholders communicated the information of climate change crisis and the issue has drawn unprecedented attention from the whole world. Compared to earlier conferences, the COP15 was the debut that the Chinese government, media and NGOs participated the international climate change conference stage.

Full of twists and turns, COP15 was full of conflicts and highly dramatic, reflecting the international climate change negotiation was complex and tough. After COP15, China has gradually upgraded climate change response to a national strategic level, and made a series of important decisions and arrangements to stepup energy saving and emissions reduction, effectively control the greenhouse gas emission to address climate change. Different stakeholders of China have consensus to cooperate at both international and domestic levels. After COP15, the government, media and NGOs have continued to follow up the negotiation, also actively carrying out the climate change response work at home. In 2015, the Paris Agreement was signed as a result of concerted efforts of different parties and stakeholders.

1 Role Changes of Chinese Government

1.1 Role of the Chinese Government in COP15

COP15 in 2009 attracted environmental ministers of 192 countries and heads or government leaders of 85 countries and surpassed any previous international talks. The Chinese government delegation in COP15 was a high-level team consisting of the representatives from a number of ministries/commissions and Chinese scientists. At that moment, the Chinese government delegation played as the negotiation entity and the main information resource.

Role 1: Negotiation entity

The UNFCCC and the Kyoto Protocol both set out the objectives and tasks for different contracting states to respond to the climate change, clarifying the status of countries as the key participants in global climate governance. Delegations participating in the international climate conference mainly consisted of government representatives. The government is the maker and implementer of domestic policies as well as the entity of international negotiation.

The Chinese Government displayed an unprecedented initiative to promote the COP15 to reach an effective international agreement. Compared with the passive defense strategy in the prior international negotiations, the Chinese Government took a different strategy during COP15.

Before attending the COP15, the Chinese Government announced the national action targets to address climate change in the following more than ten years. The Chinese Government dispatched an official delegation with more than 100 members from the National Development and Reform Commission, the Ministry of Foreign Affairs, the Ministry of Finance, the Ministry of Science and Technology, the Ministry of Ecology and Environment, China Meteorological Administration and other related ministries/commissions, and also a legal team with over 50 members from related research institutes. It's the first time that the official delegation has such a large scale.

During the COP15, the National Development and Reform Commission played the role as coordinator to collaborate with the Ministry of Foreign Affairs based on different negotiating responsibilities, assisted by China Meteorological Administration, the Ministry of Finance, the Ministry of Science and Technology, the Ministry of Ecology and Environment, the State Forestry Administration and other departments. Besides, the team comprising more than ten famous scientists in China's climate area also worked hard to express the stances of the China at the side events outside the negotiation.

During the first week of the negotiation, the Chinese Delegation explicitly pointed out that the autonomous action targets already proposed by China were nonnegotiable and defeated the intent of the developed countries to press China to undertake the responsibility and obligation in excess of its capacity, and effectively safeguarded the right and space of development of China and other developing countries.

During the second week of the negotiation, the then Premier Wen Jiabao attended the conference. Within three days, he performed frequent diplomatic activities, and met the leaders from both the developed countries and the developing countries. He actively exchanged views in depth with different parties to narrow the gap and promote the consensus. Seeing Premier Wen mediated on behalf of the Chinese central government, the Chinese Delegation became highly confident in achieving the negotiation results. The fact has proved the Chinese Government had optimistically anticipated the situation of the negotiation and underestimated the capacity of the counterparties of game.

During the negotiation, the divergence among developing countries were bigger, the BASIC countries didn't gain adequate international influence, and the international NGOs, potential to be allies, welcomed the sudden positive attitude of China. However, they, as key stakeholders, also took a prudential watch-and-see attitude and kept neutral on relevant matters. 192 contracting states participating in the negotiation generally opted to hold fast to the upper limit and squeeze the space for compromise and negotiation and refused to make any compromise in relation to their national interests. Finally, the negotiation failed to achieve an expected result but only ended with an agreement without legal binding force, which was far from enough to satisfy the expectation of the international community. Reluctant to lose the dominant right of the negotiation, the U.S. and the EU stressed that the agreement reached by China and some countries couldn't represent the opinion of all contracting states and emitted signals through international media that China played "little tricks".

The Guardian published an article at the first time, namely "How do I know China wrecked the Copenhagen deal? I was in the room", which claimed to be written by a negotiation representative. The author described the so-called "truth" he observed and blamed China for "wrecking" the Copenhagen Negotiation. The U.S. took advantage of the situation and made China the scapegoat that should be responsible for the failure to reach a legal document during Copenhagen Negotiation. The Chinese Government Delegation didn't expect such a sudden change around the ending of the negotiation, and at this time, Premier Wen, who was the core of the supreme decision-making level in the front line, already returned to China after completing the mediation. Thus, the Delegation was absent from decision-makers to some extent and couldn't respond to such sudden changes in time. The journalist of *The Guardian* also mentioned this point in an interview at the beginning of 2010, "When the Chinese Government makes a decision, it seems that all countries must agree with it. When other countries disagreed or the situation suddenly changed, they would not know how to make decisions and must report to the supreme decision maker".[1]

Constrained by the top-down decision-making system, the Chinese Delegation failed to clarify the misunderstandings of the international media at the first time but kept silent for 48 h successively, thereby missing the best time to voice out. China was then imposed a negative label of the "kidnapper of Copenhagen Negotiation" and suffered a serious impact on its international image.

Role 2. Main information resource

The Chinese Government Delegation participated in the negotiation from beginning to end. They were best qualified to speak on the negotiation process and played the role as the main information resource. The journalist of the China Dialogue recalled that during the COP13 in Bali in 2007, "China didn't make communication effort at all but only put a booth, on which there were some Chinglish brochures without any staff around. The journalist strongly asked the Delegation to hold a press release. Finally, China held a press release but excluded foreign media".[2] Compared to two years ago, the Chinese Government Delegation set up a special news and exchange center inside the venue of COP15 to hold regular press releases, exhibit the country's developmental achievements and respond to the latest progress of the negotiation.

On the afternoon of December 7, 2009, the Chinese Delegation held the first media briefing for Chinese journalists at the center. Zhenhua Xie, Head of the Chinese Delegation, reiterated the standpoints of the Chinese Government. On the afternoon of December 8, Wei Su, Deputy Head of the Delegation shared the basic progress of the negotiation in the same place and accepted an interview with Chinese and international media journalists. It's the first time that the Chinese Government Delegation organized the press release open to international media.

Confident to live up with questions from international media, the Chinese Government took a key step on the way of climate change communication. Later, the Delegation successively held irregular news releases at the center and members or experts of the Delegation communicated with media about different hotspot issues. Zhenhua Xie and Wei Su, severely criticized some developed countries for violating the UNFCCC and the Kyoto Protocol at the news releases on various occasions. Such active arrangement quickly put the center in the focus of global media.

During the COP15, the spokesperson of the Chinese Government Delegation displayed flexible method and skillful expression at the news releases, which created some influences.

For example, during the negotiation, the EU and the U.S. required China to "express more" by promising to provide some financial aids for the developing countries. At the news release on December 8, Wei Su fought back with basic common sense: The developed countries said that in 2010–2013, they would be willing to annually provide US$10 billion to help the developing countries to address climate

---

[1] Source: Interview with the stakeholders designed together with the author in 2010.
[2] Source: The interview with the journalist of the China Dialogue in the interview with the stakeholders designed together with the author in 2010.

change, but this amount, if evenly distributed to the population of the developing countries, would be only US$2 per person, "which is even not enough to buy a cup of coffee in Copenhagen.".[3]

On the morning of December 10, the U.S. representative Todd Stern refused to provide the climate financing support for China in a speech. When western media kept asking about the attitude of the Chinese Government at the press release in the afternoon of the day, Wei Su responded, "First of all, I respect Mr. Todd very much, and I and Mr. Todd are good friends as well. No single contracting state can determine whether China can obtain the climate financing support. As a contracting party, he can spend his money as he like, as long as he doesn't violate morals and provisions of UNFCCC. If we discuss this issue under the UNFCCC, the developed countries are obligated to provide financial and technical supports for the developing countries to help them mitigate and adapt to the climate change." Before ending the speech, Wei Su adjust the tone in a soft and open way, "of course, I sympathize with Todd very much for having to receive an interview with the journalists just after taking off the plane. Todd, thank you for your hard work!" Such kind of answer mitigated the intense situation on the site, which won the Chinese Delegation an affirmative applause.

Another detail must be pointed out that during the press releases on the first two days, the Chinese Government Delegation didn't publish an announcement through the conference but only sent an email to a small number of Chinese journalists and restricted the participation of international media. Later, the Delegation learnt from others and published the information at the press releases to be held at the center on the schedule column of the official website of the conference one day in advance.

However, the Chinese Government still had some obvious problems related to information release, restricted by the administrative system and thinking inertia, according to the interviews with the media participating in the negotiation, including the *21st Century Business Herald, China Daily, Caijing, First China Business News, China Dialogue* and *The Guardian*.

First, the forms of information publishing were rather rigid. Haili Cao, a journalist of *Caijing*, said, "the press release lasted a total of 30 min, but the official took 25 min to give a speech, which left very limited time for journalists to ask questions. Besides, the official used the language kind of implantation and lacked the interaction about information exchange".[4] Besides, the China news and exchange center was limited in size, which restricted the quantity of media participants. Moreover, the press release didn't take the form of Internet streaming, so it couldn't effectively convey the voice of China to the audience caring about the progress of the conference over the Internet. I saw on the site that many journalists were kept out of the center but had to put their ears on the wall or stand high and reach the recording microphone into the center to know the content of the official release. In an interview, Zhijian Zuo, an editor of the *21st Century Business Herald*, said that compared to diversified forms of information

---

[3]Source: The interview with the journalist of the *Caijing* in my interview with the stakeholders in 2010.

[4]Source: The interview with the journalist of the *Caijing* in 2010.

release taken by other countries, the Chinese Government still limited the channel of information publishing to the most traditional form of press release and opened no other platforms to communicate with journalists. "They lacked the diversified strategy and failed to do the work in great detail", and "they lacked 'Off the Record' and Background", two forms often used by the governments of other countries."[5]

Second, publishing contents were less attractive. From the perspective of two-level game, China should emphasize domestic realistic difficulties, such as poverty, ecological vulnerability and conflict between emissions reduction and development, for the purpose of seeking the right of development and the survival space in the international community. However, affected by the rigid positive communicative style, the Chinese Government Delegation avoided publishing problem and challenge information to establish a positive international image. In Copenhagen, the Chinese Government still adopted the style of press releases in China and only exhibited achievements without mentioning difficulties and problems. "This makes me feel untrue". The published contents were less attractive to international media. The journalist of *The Guardian* said, "It is very contradictive to participate in the press release of China. It is the only channel of information exchange China had opened. If I go, I will hear so many empty words, but if I don't, I will even have no chance to exchange with them".[6]

Finally, the communication plan was not adequate. The Chinese Government should be acknowledged with its performance during the first ten days. However, after suffering a misunderstanding before the end of negotiation, the Chinese Government had no plan B but just cancelled the press release already arranged. This directly muted the Chinese Government and media, which were blamed by international public opinions.

Role 3. Administrator

The government, media and NGOs are major stakeholders participating in the international negotiation, share common values and objectives under global climate governance and represent the stakeholders with positive correlation. However, during the COP15, the Chinese Government didn't see media and NGOs as stakeholders, but still adopted the domestic administrative style without any adjustment: The government strictly controlled media coverage and didn't pay attention to the existence of NGOs.

When cooperating with domestic media, the Chinese Government Delegation managed contents of media coverage by distributing news releases and arranging reporting agenda at key time nodes. These methods have the advantage of helping a lot those Chinese media not familiar with the climate change negotiation to follow the negotiation progress and avoid serious reporting efforts. However, they have

---

[5]Source: The interview with the journalist of the *21st Century Business Herald* in my interview with the stakeholders in 2010. "Off the record" means an interview without recording and can provide journalists with some insider information, and the journalists will not designate the information source. "Background" means inserting the interview material as background information into the news release.

[6]Source: The interview with the journalist of *The Guardian* in 2010.

also led to homogeneity, lack of profound analysis and other problems, which make media incomparable to international counterparts in terms of quality. In particular, when facing the challenge from international media around the end of the negotiation, Chinese media, which were used to relying on the government as the main information source, could do nothing but keep silent together. *The Guardian's* article directly attributed the failure of the COP15 to China. Such voice went beyond the expectation of the Chinese Delegation. China kept no internal urgent response mechanism and also lacked the channels of information exchange with other stakeholders. As a result, the government couldn't set the tone for the reports of Chinese media as before, distribute a press release to media in time or prompt them of key points. When the government instruction was absent, the Chinese media collectively kept silent in face of the challenge. In the eyes of the international community, China acknowledged the report of *The Guardian*. Two days later, the Xinhua News Agency was authorized to publish a signed article entitled "The green mountain can't stop the river, which will flow eastward after all—Records of Premier Wen Jiabao's Participation in the Copenhagen Climate Change Conference". However, this already missed the best timing to respond to the challenge and didn't work a lot to reverse the negative opinion of the international community about China. Moreover, the title of the article quoted a Chinese ancient poem which means the current obstacle cannot prevent the long-term development, but the international community couldn't understand the profound meaning behind it. Besides, the writing style was outdated, so the conveyed message couldn't be accurately accepted by the international community to change the game.

As to cooperation with NGOs, the Chinese Government Delegation didn't realize the rising role of NGOs in global governance during the COP15 or research their contribution and role in the international community in depth. Instead, the Delegation opted to shield relevant messages from NGOs and avoided any active contact, which made it impossible to coordinate and interact at critical moment.

In conclusion, the Delegation had no problem in relation to negotiation entity and positioning as information source. However, the role of the top-down administrator restricted the performance of the Chinese Government on the international stage and also hindered cooperation among stakeholders, which plunged China into a negative situation.

Hindered by the then domestic decision-making mechanism and thinking inertia, the decision authorization mechanism was not flexible, and the government didn't understand counterparties of game in the international community or perform the scenario analysis of negotiation results from multiple angles.

From the perspective of two-level game, the domestic game and the international game have respective characteristics, and corresponding strategies should be differentiated to avoid blind copy or duplication. The domestic administrative practices of the Chinese Government didn't apply to the international climate change negotiation, and the domestic strategy shouldn't be directly applied to the international game without change.

1.2 Tracking Analysis: Changes in Government Strategies at International Level

The lessons learned during the COP15 stimulated the Chinese Government to review the negotiation and cooperation strategies in depth. From Copenhagen to Paris, the government has remained the roles as negotiator and information resource, but obviously changed its strategies. The government has also adjusted the role of "administrator" and corresponding strategy after review.

1. **Flexible, open negotiation entities**

The UNFCCC stresses that the developed countries should assume major historical responsibility for the climate change and take the lead to perform the obligation of emissions reduction. At the current stage, the developing countries should mainly pursue development and can perform voluntary emissions reduction based on respective capacity. The principle of "common but differentiated responsibility" stipulated by the UNFCCC is the baseline the Chinese Government has adhered to during participation in the international climate change negotiation over the past six years, that is, require the developed countries take the lead to reduce the emission of greenhouse gases and China will not warrant mandatory emissions reduction before becoming a moderately developed country. While holding fast to the baseline, the Chinese Government became more flexible and open towards the negotiation.

First, seek the support from key counterparties of game.

During Copenhagen Negotiation, the Chinese Government maintained normal communication with various counterparties of game. Premier Wen Jiabao arrived at the site in the second week to mediate with other countries, which was a historic breakthrough. However, the objects of mediation were only developing countries, and China didn't pay much attention to communicating with the U.S., the EU and other key counterparties of game with bigger influence and dominant right. As a result, China suffered misunderstanding and blame when the negotiation ended. After learning the lesson from Copenhagen Negotiation, the Chinese Government started carefully studying the standpoints and strategies of different counterparties and strived to establish strategic partnership with key counterparties of game to reach the win-win targets. The most typical example is the cooperation between China and U.S. on climate after COP15. As one of the key steps, China and the U.S. jointly published the China-U.S. Joint Statement on Climate Change on November 12, 2014 in Beijing with the chance when the U.S. President Obama visited China. With this statement, both countries announced respective emissions reduction plans. The U.S. has promised to double the speed of carbon emissions reduction from 1.2% to 2.3%–2.8% annually on average after 2020 and reduce the carbon emissions by 26%–28% by 2030 compared to 2005. China has declared to reach the carbon emission peak around 2030 and gradually reduce the emission annually afterwards. According to the plan, the proportion of non-fossil fuels in China's energy structure will reach 20% in 2030. The joint statement is the first time that the heads of China and the U.S. have declared their action plans for climate change response after 2020. China and the U.S.

contribute nearly half of the world's total carbon emission and energy consumption, one third of the world's economic aggregate, one quarter of the world's population and one fifth of the world's trade value. The joint statement of both countries has set a very good example promoting global emissions reduction and played an essential role in propelling the progress of the international climate mechanism.

Second, abandon zero-sum game and create the space for cooperation.

The zero-sum game means that during the game, if one party wins, the other party will definitely lose, and the result of neutralization between both parties will be zero and contain no possibility to cooperate and achieve a win-win situation. The global climate governance is the common governance performed for common interests of mankind, so it stresses cooperation and mutual benefits and is typically not a zero-sum game. If all parties in the game refuse to yield and vow to win for their interests, the international climate mechanism will not make effective progress. The global climate governance will require different parties to suffer certain sacrifice in the short run but will benefit all parties in the long run. After realizing this point, the Chinese Government gradually abandoned the concept of zero-sum game and actively created the space for cooperation.

The Durban Platform for Enhanced Action built in 2011 is a monorail negotiation. The Chinese Government always adhered to the "dual-track" negotiation but also actively propelled the landing of the Durban Platform on the principle of "common but differentiated responsibility" under the Convention. For example, as to the financing and technical issue during the negotiation, China initially insisted the developed countries should provide financial and technical aids for the developing countries. Yet, after further research, China started calling for the establishment of the technical promotion mechanism and technical cooperation mechanism that would benefit both parties. China also set an example by actively carrying forward the south-south cooperation in response to climate change, providing financial, technical, product and training supports for other developing countries and building the south-south cooperation platform for climate change. In the second China-U.S. joint statement published in November 2015, China increased the financial support for south-south cooperation to US$3.1 billion. When addressing the opening ceremony of the COP21 in Paris in 2015, Chinese President Xi Jinping convened the standpoint of abandoning the zero-sum game to the international community and further explained concrete use of US$3.1 billion fund, which helped China win a positive position during the negotiation on the funding issue in the first week.

Finally, China actively transformed the game strategy.

Mitigation and adaptation are two core issues of climate change. To respond to the climate change challenge, the mitigation of greenhouse gas emission and adaptation to climate change should be equally important. Adaptation to climate change will be more important than mitigation for the developing countries affected by climate change, which is also the domestic root cause for China's participation in international climate change negotiation. The international climate mechanism, initiated by the developed countries, always pays more attention to emissions reduction than adaptation. After the Copenhagen Negotiation, China has changed the negative tracking of negotiation issues and changed from pure emphasis of emissions reduction to simultaneous emphasis of relevant issues, including mitigation and adaptation.

2. **Information publishers with diversified strategies**

First, the publishing channels became more diversified and forms became more flexible.

During the Copenhagen Negotiation, the Chinese Government only opened press release as the traditional publishing channel, and speeches of the spokesperson usually adopted the implanting language, which restricted the effect of information communication to some extent. After reviewing the lesson from the Copenhagen Negotiation in earnest, the Chinese Government became more flexible in channel of information release and form of release.

Since 2011, the Chinese Government has opened the China Pavilion inside the venue of the COP as a fixed platform of information release. In the year, China organized 23 serial events and exhibited the efforts of China in relation to low-carbon development, climate financing, corporate action, civil energy saving and other aspects through forum, press release, seminar, exhibition and other forms. The representatives participating in these events came from governments at various levels, NGOs, enterprises and research institutes, many international experts also came onto the stage to give speeches, and the events attracted attention from negotiation representatives and journalists with rich contents. To track China's latest policies or actions against climate change, international journalists would only need to watch the schedule distributed around the China Pavilion to participate in relevant events in time.

Reviewing the themes of the series of side events in the China Pavilion, we can find that the themes have grown richer and richer amid the government's deepening understanding of climate change. Since 2011, "climate change communication and public participation' has become one of the regular themes of side event. In 2014, the side event of the China Pavilion themed "climate financing" was held for the first time. In 2015, the China Pavilion first held a side event on "climate change and rural development" since its opening 4 years ago. In this sense, we can see that with the China Pavilion as a window, the Chinese Government has demonstrated to the international community that it has deepened the understanding of climate change.

The Chinese Government also attempted to build the China Pavilion as the platform for communicating China's soft powers, besides defining it as a formal information release platform. the China Pavilion prepared Chinese fans and other personalized gifts in Doha in 2012, and directly borrowed ancient, simple and beautiful Chinese elements, including Anhui-style architecture, water-and-ink landscape drawing and green bamboo of southern China, in decorative style during the conference in Lima, Peru in 2014. In Lima, the China Pavilion also specially designed an ancient lobby, an area where representatives of different countries visiting the corner could take a rest and experience Chinese cultures. The embedding of these Chinese elements injected a fresh style into the venue of COPs and also strengthened the affinity of the Chinese Government.

Besides, the Chinese Government also paid more attention to diversifying forms of release. During the Copenhagen Negotiation, the Chinese Government released information mainly through traditional channels such as press conference and news release, and government officials were reluctant to talk a lot in private. After reviewing the lesson in Copenhagen Negotiation, the Chinese Government Delegation displayed obvious changes in this regard. For example, the offices of the Delegation opened wide to journalists to enable more direct communication. When having any question, journalists could go to the offices anytime to know the government's attitude. Scientists clearly defined their role as popularizing professional knowledge involved in the negotiation and answering questions of journalists. They were prepared to accept interviews with journalists and even took the initiative to clarify relevant questions to media.

Second, speak in the right timing.

The negative result during Copenhagen Negotiation made the Chinese Government realize the importance to speak in time. During the Durban negotiation in 2011, the international public opinions were full of rumors, including the BASIC countries "were at the verge of separation" or "had serious debate". On December 6, the BASIC countries held a ministerial press conference, and China actively responded, "the BASIC countries firmly unite with one another. We are all responsible in response to climate change and taking positive actions. Moreover, we have already made some progress" (Yu 2011). Before the Durban negotiation ended, the negotiation tended to crash. Zhenhua Xie, Head of the Chinese Government Delegation, strongly blamed the developed countries for failing to perform their due obligations when addressing the conference, "for some countries, we don't hear what you say, but watch what you do. Some countries have made commitments but have not delivered their commitments or taken substantial actions." Later, Zhenhua Xie gradually raised his voice and kept asking questions, "You have promised you would take the lead to perform substantial emissions reduction. Have you done this? You have promised you would provide financing and technologies for the developing countries. Have you done this? You have kept saying that for 20 years but have not delivered them so far. China is a developing country, and we have to pursue development, eliminate poverty and protect the environment. We have done all we should do but you have not done what we have done. Then, what has empowered you to gossip about such things?" The speech of Zhenhua Xie has won unanimous compliment from the representatives of the developing countries and his speech video has widely spread on video websites.

Since 2013, a crowd of young negotiation representatives have grown up, and the first-generation representatives headed by Wei Su have gradually gone behind the curtain. Young negotiation representatives all speak fluent English and speak on the negotiation site in time. Instructed by senior representatives in content, they have further strengthened China's capacity to voice out timely.

3. **Administrator attempting cooperation**

People of NGOs, particularly international NGOs, are full of enthusiasm and innovation. Moreover, with technologies, knowledge and expertise, they can convey China's attitudes and behaviors to the international public in a casual and friendly manner, promote exchanges between China and the world, thereby becoming the major drive for public diplomacy.

Before Copenhagen Negotiation, the government didn't pay adequate attention to NGOs, particularly international NGOs, as key stakeholders. After the COP15, China Center of Climate Change Communication (China4C) was set up by some witnesses of COP15, as the local think tank, to conduct research on the cooperative role of key stakeholders and shared the key findings to the Chinese government. To some extent, The COP15 in Copenhagen has enabled the Chinese government to notice the important role of international NGOs and the Chinese Government started attempting to cooperate with NGOs.

The first attempt happened during the UN climate conference in Tianjin in October 2010. During the break, Minister Zhenhua Xie, Head of the Chinese Delegation, visited the booths of the NGOs outside the venue and met the representatives from 21 NGOs. He acknowledged their role and exchanged views with them on the negotiation agendas.

Afterwards, similar communication has become normal. Before every negotiation, Minister Xie will invite NGOs and media to meet respectively and exchange views with them in relation to China's standpoint, analysis of international public opinions, forecast of negotiation status and other issues. From the Durban negotiation, the Chinese Government Delegation has assigned representatives to attend side events hosted by NGOs. For example, Wei Su, the chief negotiator of the Chinese Government Delegation, delivered keynote speeches at the side event on climate change communication and public participation co-hosted by the international NGO Oxfam and the Renmin University of China. The NGOs were also invited to attend a series of side events at the China Pavilion. From 2012 to 2014, the China Pavilion organized side events for NGOs in succession. China's local NGOs and international NGOs specializing in climate change in China briefed their work at these meetings and exchanged with foreign counterparts and government representatives. In 2015, the list of honorable speakers at different side events of the China Pavilion included the representatives from NGOs. This proves that the government has established basic consensus on climate change with NGOs.

1.3 Tracking Analysis: Changes in Government Strategies at Domestic Level

When judging whether the international climate governance is internalized, we can use four indicators, including perception change, system reform, legislative support and policy practice. The pressure China has felt at the international level over the past six years has become the motive to carry forward climate change response at the domestic level, and the international climate system has been internalized to a big extent in China. Accordingly, China has also changed from a negative participant to a positive partner and the Chinese Government has become the advocate and disseminator of international climate governance mechanism.

1. **Upgrade of climate change response as a national strategy**

The Chinese Government has gradually paid rising attention to climate change response and experienced the change from perception to action. We can see from the interviews with government officials in 2010 that a dispute in the understanding of climate change even existed within the central government before Copenhagen Negotiation. Some government officials believed that "climate change is a hoax the western countries had devised to restrict China's development". Some officials even contended that the "low-carbon economy is a measure that doesn't suit China's situations".[7]

With global rising attention to climate change and China's expansion of economic strengths, the central government has gradually realized the disadvantages of the extensive development pattern and the realistic pressure from resources and environment. To attain the strategic objective of sustainable development, China should transform the development pattern and realize the transformation towards low carbon. Low carbon is a definition relative to "high carbon" and also a new term that has appeared in the context of climate change. Transformation towards low carbon demands China abandon the former production mode of high emission, high pollution and high energy consumption and select new energy and renewable energy as substitute. Such transformation will help China fulfill the objective of sustainable development and also help China play a constructive role in global climate governance. Driven by the consensus within the high-level leadership, China has written climate change into many documents issued by the central government and called upon the public to pay attention to low carbon and emissions reduction through the top-down communication.

To address climate change, we should simultaneously pay attention to mitigation and adaptation. For the developing countries, adaptation is more imperative than mitigation. The popularity of knowledge about climate change and the increase of China's scientific research level have frequently proved the fact that China is affected by climate change. It has been put on the agenda to compile the National Climate Change Response Strategy. After two years of refinement, China issued the first National Climate Change Adaptation Strategy in 2013. Adaptation to climate change has been included in the 12th Five-year Plan, which reflects China attaches much importance to adaptation.

In 2014, the Chinese Government further upgraded its attention to climate change. China decreed the National Climate Change Planning (2014–2020), which elaborates the strategic significance of climate change response. The China-U.S. Joint Statement

---

[7]Source: The interview with the government representatives in the interview with the stakeholders designed together with the author in 2010.

on Climate Change published in November 2014 is a material symbolic result that China and the U.S. have achieved in the area of climate. In this statement, climate change is no longer "one of the biggest challenges facing mankind" but described as the "biggest threat facing mankind", thereby upgrading the importance and urgency of climate change to the supreme level. Besides, the joint statement also associates climate change with national security and international security and demonstrates climate change response already has become strategically important.

2. **Supporting media**

Just as analyzed earlier, the Chinese Government continued adopting the domestic style and implemented the parent-like style of media management during the Copenhagen Negotiation. The Chinese Government also observed that if media were allowed certain flexible space, they would play a positive role in propelling the negotiation, which has also promoted the Chinese Government to reflect its way to deal with media. When receiving interviews after the end of the Copenhagen Negotiation, some representatives reflected that the government "should allow media to exert their occupational sensitivity and seize the best communication time" to avoid suffering the negative situation during the negotiation again.[8]

Over the past years, the Chinese Government has become more and more flexible to deal with media affairs and gained richer and richer experience. Now, it can use data and facts to prove its opinion and standpoint. To make journalists more professional, relevant government authorities have regularly organized related theme trainings. In 2014, the National Development and Reform Commission, the State Forestry Administration and China Meteorological Administration jointly organized four media trainings in collaboration with China4C. These four events were all arranged before important events related to climate change, such as the Tree Planting Festival in April, China's Low Carbon Day in June, the New York Summit in September and the COP20 in Lima in December to help journalists clarify the reporting concept. Such arrangement can provide more relevant reporting materials for journalists in the short run, and can improve their professional level in the long term.

Besides, related government authorities have taken the initiative to organize field interviews for media and help them understand the impact of climate change on China. The "Responding to Climate Change • Recording China" event is a field survey and scientific communication campaign initiated by China Meteorological Administration. Since 2010, the campaign has left footprints in the Three Rivers' Source of Qinghai Province, Alashan League of Inner Mongolia, Poyang Lake of Jiangxi Province, Hongshui River Basin of Guangxi Region, the coastal cities of Guangdong Province, the Dongting Lake of Hunan Province, Xilin Gol of Inner Mongolia, Hexi Corridor of Gansu Province, Qinling Mountain of Shaanxi Province and other regions. The survey team has visited the typical areas affected by climate change response, validated years-long observation and research results in this regard from the perspective of meteorological science research and media communication.

---

[8]Source: The interview with the government representatives in 2010.

3. Cooperating with NGOs to motivate public participation

Based on the two-level game theory, the international level and the domestic level influence one another. At the international level, the Chinese Government has perceived the role of international NGOs as key stakeholder and started exploring the way of cooperation from the Tianjin conference. From rising exchanges, the government has deepened the understanding of NGOs and the latter have better understood China's actual challenges. In particular, with the rise of China, the government has refreshed the perception about the importance of global governance and accepted sustainable development as an objective. After years-long mutual adaptation, the international NGOs and the Chinese Government have changed from potential counterparties of game to partners sharing the common objective, guided by common objectives.

As understanding the role of NGOs in greater depth at the international level, the government has also gradually started to cooperate with NGOs at the domestic level.

NGOs have natural advantages in mobilizing public participation. From the perspective of mission, NGOs work to power a sustainable development of mankind. From the perspective of working methods, NGOs mainly attain objectives by mobilizing the public and creating public opinions. Climate change response calls for the public participation and action, and the precondition of action is perception. Since 2013, China has specially set the National Low Carbon Day and organized various themed campaigns together with the World Wide Fund for Nature (WWF), the Environmental Defense Fund (EDF), the Friends of Nature and other NGOs to make the public more aware of climate change and low-carbon development, encourage more public participation and action and attain the emissions reduction task. Over the past two years, the objectives of the National Low Carbon Day have gradually shifted to public participation as the public is more aware of climate change. With the theme of "Low-carbon City, Livable and Sustainable", the National Low Carbon Day in 2015 launched a national communicative campaign and cooperated with the international NGOs and local NGOs to provide operable action suggestions.

Considering the natural contact of NGOs with the public and their network advantages, the government has invited the representatives of NGOs to participate in rounds of discussion and provide policy suggestions when carrying forward the legislation regarding climate change.

2 Role Changes of Chinese Media

The two-level game theory stresses the decision-maker faces both the international and domestic fields of game, and media play an essential role to convey information during the interaction. At the international level, media shoulder the mission to convey the message and story about China's actions timely and accurately, reduce the misunderstanding caused by poor communication and promote China to make more constructive contributions to global climate governance.

At the domestic level, media should moderately convey the pressure faced by the Chinese Government on the international field of game, communicate the urgency, importance and concrete action points of climate change response and boost the internalization of the international climate mechanism. At the same time, they can enhance the public awareness, motivate the public to take more tangible climate actions, promote China to form the largest set of winners and win more weight for China to participate in the international negotiation.

2.1 Role of the Chinese Media in COP15

Hitting a new historic high in importance of agenda, size and level, the COP15 has attracted nearly 10,000 journalists from across the world to assemble in Copenhagen. Half of the journalists completed successful registration and entered the venue to report the negotiation progress, and the other half wandered on the streets of Copenhagen to capture the information in relation to climate change and negotiation. Great attention and frequent reporting of media have made the negotiation and the city of Copenhagen come into focus of global public. The performance of media also became one of the focuses in the public opinions of the society. Most of Chinese media first participated in the work to report the international climate change negotiation. For this reason, they lacked professional knowledge and barely played the role of communicating information. In comparison, international media were more flexible in role positioning and communication strategy, and boosted the negotiation.

Role 1. Communicator

The communicator of information is the basic role definition of media at the international and domestic levels. Most of Chinese media first participated in the work to report the international climate change negotiation. To play the role well, they made preparations to varying degrees before start. For example, they learned scientific knowledge related to climate change, reviewed the history of climate change negotiation, agendas of negotiation and China's major counterparties. *Caijing* and the *21st Century Business Herald* made better preparations.

For example, the front-line reporting team of *Caijing* consisted of four journalists. Three journalists followed up the latest news published by the U.S. Delegation, the European Delegations, the Chinese Delegation and the NGOs, and the other was responsible for random interview outside the venue. Judging from the division of work, they basically could follow up speeches of key negotiation representatives and flexibly captured breaking news. Besides, to offset the low timeliness as a magazine, *Caijing* also cooperated with Tencent.com: While writing features and accumulating materials for the magazine, the *Caijing* journalists also furnished the news with strong timeliness to Tencent.com and published such news on the Internet platform.

Considering the time difference of several hours between Copenhagen and Beijing, the reporting team sent by the *21st Century Business Herald* also included one editor, besides four journalists. The editor worked in the news center of the venue, treated the articles completed by the journalists anytime and contacted with the domestic team to assure the timeliness of news. At the same time, the magazine already adopted the editor's responsibility system, whereby the editor allocated topics to the journalist, at that time. The editor working at the frontline could assure the agenda setting and allocation of topics were reasonable".[9]

Besides, as I observed on the site, the Xinhua News Agency, the *First China Business News* and *China Daily* also made corresponding preparations. However, compared to international counterparts who have tracked the climate change negotiation issues since the 1980s, only a few journalists of Chinese media participated in the Bali negotiation in 2007. Most of the journalists didn't pay much attention to the climate change issue and lacked the long-term accumulation of professional knowledge on climate change negotiation, so there were no professional journalists in the area.

The preparations made in advance were far away from enough to satisfy actual needs. Judging from the understanding of negotiation agendas, the journalists became familiar with possible contents of the negotiation, but most of Chinese media could only interpret the agendas in single dimension and lacked the ability to understand and parse these agendas in depth, because the issue of international climate change negotiation was historically continuous, extensive and systematic. Judging from the agenda setting, Chinese media lacked the capacity to independently judge the negotiation progress and excessively depended on the guidance of the government Delegation, as a result of inadequate accumulation and preparation. Regarding the reporting content, Chinese media collected a large number of trailers outside the venue because they couldn't follow up the negotiation from the professional perspective. From the reporting audience perspective, the media didn't distinguish contents of the reports oriented to the needs of the audience at the international level and the domestic level, and instead, the contents were highly homogenous and incapable to attain the objectives of climate change communication at two levels. For the aforesaid reason, Chinese media became silent collectively, when the Chinese Government Delegation failed to respond to the international media after incurring a problem with the emergency response mechanism. As the communicator of information, Chinese media suffered a negative situation and failed to create a big international influence and contribute to helping China establish the positive image and conveying the international pressure.

Role 2. Negotiation booster

The Copenhagen Negotiation had two key nodes: First, Denmark's secret document was divulged in advance, and second, China, who conducted active mediation, was blamed as the kidnapper of the negotiation. The explosion of both nodes was closely associated with *The Guardian*. The Danish document was a document prepared by the host country Denmark in advance in the hope of communicating details of the agreement and promoting the signing of an international agreement through the

---

[9]Source: The interview with the journalist of the *21st Century Business Herald* in the stakeholders designed together with the author in 2010.

consultation off the negotiating table. The core content of the document is to allocate the emissions reduction indicators and require the developing countries assume the mandatory obligation of emissions reduction, which doesn't observe the principle of "common but differentiated responsibility". The Danish document was divulged to *The Guardian* journalist in the second week of the negotiation and triggered strong dissatisfaction of the developing countries after it went public. Due to the public opinion pressure, Denmark had to cancel the discussion about its document. The report of *The Guardian* helped the developing countries avoid a negative status to some extent. While right after the negotiation ended, it's *The Guardian* that published a signed article at the first time and shifted the international public opinion pressure to China.

Obviously, the journalists of *The Guardian* have rich accumulation, high professional level and quick response in the reporting of climate change negotiation, and own the ability of agenda setting to boost the negotiation. If it is the basic role of the media to timely report negotiation progress and take advantage of the attention to the negotiation to disseminate the knowledge on climate change, then, they can further play the role as negotiation booster on the basis of professionalism and independence. Yet, Chinese media are still relatively strange with this role and need to accumulate and learn more.

2.2 Tracking Analysis: Changes in Media Strategy at International Level

Over six years after the end of the Copenhagen Negotiation, Chinese media have obviously improved the quality of their climate coverage. They have further defined the role as climate communicator, adjusted corresponding strategy and consciously learned from international counterparts to move closer to the role of negotiation booster.

1. **Flexible strategy for communicator**

First, preparations are more complete. Before Copenhagen Negotiation, the preparations of media were limited to supplementing knowledge on the agendas. With the lesson in Copenhagen, media have realized the importance to make preparations in earnest. Besides the training organized by the government before start, Sina, Sohu, Tencent and other local online media built platforms in turn and collaborated with three major stakeholders to discuss the preparation strategy.

Second, the preparations are more tridimensional. In 2015, NetEase, one of the then most influential online media in China, organized a media team, namely "with attitude", consisting of more than 20 members to participate in the UN's climate summit in New York in September. The team became familiar with the history, form and content of the UNFCCC and COP to warm up for the COP21 in Paris at the end of the year. During the COP21, NetEase and Youthink, a local Youth NGO, cooperated and led China Youth Delegation to report the negotiation on the site. Moreover, they also organized the Youth Delegation to deliver vivid speeches to incumbent students of Paris Institute of Political Studies for the purpose of communicating climate change and enhancing perception.

Finally, the positioning became even clearer. With the experience and lesson from Copenhagen Negotiation, Chinese media decide whether and how to participate in the subsequent climate change negotiation based on their own conditions. Over the past years, the structure of China's media delegation has remained relatively fixed. Besides official media invited by the government, other major members include market media, local media and online media that register themselves, which present a rising proportion and remain stable at 15 in total.

The Chinese Government has become more flexible in media cooperation and adjusted the form and content of information release. For this reason, Chinese media have become more capable to set agendas after mastering the basic knowledge related to climate change and negotiation.

At the international level, there are mainly four media agencies faced by the international audience and publishing in English, including Xinhua News Agency, *China Daily*, China News Service and CCTV International Channel. During the climate change negotiation, the developing countries, including China, are very hard to get the chance of fair voices. For one thing, this is because international media don't know about China in comprehensive aspects. For example, some scholars have followed up some international magazines' cover stories about China and found more than 90% involve negative coverage. Chinese media should actively voice out to share the stories of China and help China better participate in climate governance. After realizing the aforesaid problems, the media attempted to become more flexible during climate change communication and collect information from multiple channels. For example, they invite international NGOs to make comments as independent third parties and present different voices in coverage. The international NGOs specializing in climate change in China acknowledge China's contributions to climate change response and also are clear about why the international community misunderstands China. Thus, their voice will make a better effect than the Chinese Government to clarify misunderstandings.

At the domestic level, Chinese media have consciously adjusted the reporting strategy and paid attention to transmitting the international pressure to China and propelling the progress of domestic climate change response in depth after Copenhagen Negotiation. After reviewing the reports on climate change negotiation over past six years, we can find media have obviously reduced trailer contents. Instead, they have increased the proportion of professional analysis and comment on climate change negotiation and paid attention to combining them with domestic policy trend. Even the online media have paid more attention to content integration.

2. **Enhancement of booster capacity**

During Copenhagen Negotiation, Chinese media were still incompetent for the role as booster. In the years to follow, Xinhua News Agency, *China Daily* and China

News Service responsible for international communication have taken the initiative to learn from international counterparts and propelled the negotiation to evolve in the direction of fairness and equality based on the principle of objective and authentic reporting.

To this point, *China Daily* has performed even better in the past years. During the Doha negotiation in 2012, several reports of Lan Lan, the journalist of *China Daily*, were reprinted by the UNFCCC's official website, which conveyed the voice of China during the negotiation with the platform with greater influence. On May 21, 2015, *China Daily* was invited to join the International Climate Change Reporting Alliance and become a founding member. The alliance was launched upon the initiative of the chief editors of *The Guardian, El País* of Spain and *Clarín* of Argentina. 25 international media worldwide were invited to become the founding members, open respective news coverage resources in related areas and perform adequate, objective and equitable coverage to make the international negotiation obtain positive and balanced results. *China Daily* is the first Chinese media invited to join the alliance, signifying its performance has been acknowledged by international counterparts.

Despite obvious progress, Chinese media still should further improve the overall level of covering the international climate change negotiation, particularly improve the professional level. An outstanding problem is the high turnover rate of traditional media staff, which has affected the content depth of China's climate change communication to some extent.

2.3 Tracking Analysis: Changes in Media Strategy at Domestic Level

The Copenhagen Negotiation is a watershed for Chinese media to cover the climate change negotiation at both international level and domestic level. Before 2009, climate change was not upgraded to an important agenda of the government, and the government didn't make many policies against climate change. Thus, media journalists lacked rational understanding of climate change and limited their coverage to trailers or further processed Chinese translations of foreign reports. After 2009, Chinese media have changed their reporting strategy at the domestic level.

1. **More objective reporting**

Before Copenhagen Negotiation, media had no professional experience in reporting climate change at the domestic level, which led to small quantity of reports, uncertain standpoint and strong volatility. A typical example is to exaggerate the impact of natural disasters from climate change or use the expression of terrible appeals for the purpose of catching eyeballs. To some extent, Copenhagen Negotiation has become the quick learning class for Chinese journalists, which have become more objective in reporting of climate change issues.

This shift in the domestic media has been evident in two emergencies of IPCC. In November 2009, as a representative institution of global climate change research, the Climate Research Center of the University of East Anglia was attacked by hackers, and emails and documents related to climate change research were stolen and posted on the website of meteorologists. The exposed contents show that climate data may be modified by scientists to support the argument of global warming. The research center is the data provider for the IPCC Fourth Assessment Report and is directly related to the development of the global climate change policy. This accident, called as "Climate Gate", has revived the skepticism of climate change. After the incident, the British side conducted a serious investigation and pointed out that the contents exposed were not enough to overturn the conclusion of global warming. Data from the National Centers for Environmental Information and the Goddard Institute for Space Studies further supported the conclusion that the global surface temperature has increased over the past 100 years. For this incident, the domestic media mostly forwarded the reports of the international media with the mentality of "looking at joke". All the media has supported the skepticism of climate change and almost no media interviewed the supporters of climate change.

After the Copenhagen Negotiation, the Chinese government and the media reached a consensus on the authenticity of climate change. The level of reporting the domestic climate change by the Chinese media has also improved. In the early 2010, the British media revealed that the relevant data on climate change threats to the Amazon rainforest mentioned in the IPCC Fourth Assessment Report came from reports that were not officially published. After the "Climate Gate", IPCC was again trapped in the "Amazon Gate". This time, the reports of the domestic media have become rational from blind. *Southern Weekend* systematically analyzed several incidents and the game behind it in its special article that "IPCC Suffers Continuous Trust Crisis" and "Is Trapped in the Amazon Gate after the Glacier Gate", allowing the domestic audience to have the opportunity to hear the responses of parties concerned to the incident so as to get their own judgments.

2. Multiple choices of discourse framework

My interviews with the journalists reveal that most media have similar confusion about climate change reports before the Copenhagen Negotiation, such as how to directly combine climate change with life experiences and what framework can help the public to have more accurate cognition of climate change.

Before the Copenhagen Negotiation, the media climate reports had fewer choices of discourse frameworks and used relatively simple ones such as social responsibility, biological extinction and disaster horror.

The social responsibility framework emphasizes that climate change is caused by human activities and are a common challenge facing by all mankind and all people have the responsibility to act and participate in the response. The positive perspective of the social responsibility framework lies on directly revealing the causes of climate change, emphasizing the impact of human actions on the climate system and strengthening the public's perception in this regard. While the social responsibility framework also has obvious drawbacks, such as slogan-style propaganda, and is unable to give specific suggestions for action after mobilizing the public's passion and enthusiasm. Repeated use of this discourse framework makes it easy for the public to have psychological conflicts and numbness and is not conducive to the carrying-out of the follow-up work.

The species extinction framework emphasizes that climate change can lead to the extinction of polar bears and other species. This framework is designed by the environmental protection organizations and introduces into the cute image of polar bears to bring the public closer to climate science. The problem of this framework is that it can only mobilize the enthusiasm of those who have special feelings for polar bears or who is enthusiastic about ecological protection. For most of the public, the extreme warming and the extinction of polar bears are far away from their daily lives, so the sense of urgency is relatively low.

The disaster terror framework emphasizes that climate change triggers the frequent occurrence of extreme weather and climate events and everyone living on the planet is deeply affected. The most successful case of using this framework is the Hollywood movie "The Day After Tomorrow" in 2004. This framework can immediately attract the public attention and has high urgency. But like the previous two frameworks, the disaster terror framework cannot provide points of placing actions for the public. Before the negotiations in Copenhagen, because of limited awareness of climate change, the media was more random in selecting frameworks, lacked independent judgment and planning, and was more common to apply a single framework directly.

In the past few years, the media has actively tried more discourse frameworks in the domestic climate change reports to bring climate change closer to the general public. At the same time, it also paid more attention to the combination of multiple discourse frameworks so as to mobilize more people to participate in actions against climate change.

The first type of discourse framework is the public health framework. In the year around 2013, China's air pollution has been severe as all localities have suffered the most serious air pollution on record and the smog index has seriously exceeded the health level. According to statistics, air pollution causes premature deaths of more than 1 million people a year and leads to expensive environmental damage. The effect of air pollution on health has become a topic of public concern. Although smog is not directly caused by climate change, the two problems have the synergetic relationship. Reducing environmental pollution and improving air quality is strongly coordinating with addressing climate change. Improving the environmental quality and reducing greenhouse gas emissions are two concepts, but the solutions and efforts are in the same direction. The media uses the public health framework when reporting smog, encouraging the public to protect the environment from their own actions while providing anti-smog suggestions.

The second type of discourse framework is the low carbon framework. Low carbon refers to lower greenhouse gases, especially carbon dioxide emissions. The emergence of the low-carbon framework is closely related to the Chinese government's emphasis on low-carbon development. Recognizing the shortcomings of the extensive economic mode, the Chinese government is gradually transforming into a low-carbon mode based on low energy consumption, low pollution, and low emissions. The path of low-carbon development has led to a series of low carbon-related new frameworks such as low-carbon transportation, low-carbon life, low-carbon economy, and low-carbon travel. The remarkable contribution of the low carbon framework series is to provide different options for climate actions. For example, low carbon can be linked to the lives of ordinary people. People can pay attention to energy saving, gas saving and recycling in their daily lives. Without reducing the quality of life, they can save electricity and gas costs to contribute to energy conservation and emissions reduction. Choosing the low-carbon life is a life attitude, and with the encouragement of the government, it has gradually become a popular way of life. In March 2013, the State Grid issued the guidance on the deployment of distributed power sources, encouraging people to install photovoltaic solar panels on the roof of their own homes, which not only can meet the daily electricity needs of their own homes, but also can sell the surplus electricity to the grid of the country to enjoy subsidies. On April 27, 2013, China News took the lead in issuing a report titled "Hebei Residential Photovoltaic Power Generation Successfully Combined into the National Grid for the First Time", doing an economic account for everyone, encouraging more families to install distributed photovoltaic panels and combining it into the national grid. Since then, the media across the country has popularized this policy opportunity by reporting pioneers of photovoltaic power generation in various places, setting an example for the policy landing, and helping more people to enjoy the policy dividend while joining the emissions reduction action. It can be seen that the series of discourse frameworks related to low carbon have many advantages in communication, not only are closely related to climate change and conform to the policy direction, but also connect with the traditional virtues of China, and can mobilize the interest of the public and bring real benefits to the public.

In the use of frameworks, the media also combines with the characteristics of different discourse frameworks and tries to adopt the strategy of using multiple frameworks together. For example, the social responsibility framework is used to establish a conviction for the public, the disaster terror framework is used to enhance the urgency of addressing climate change, and then the low-carbon economic framework is used to propose specific points of placing actions. The use of the strategy of portfolio frameworks has played a positive role in achieving a shift from cognition to action in addressing climate change. In November 2012, the large-scale 11-episode documentary "Warm and Cold We Share Together" was broadcasted at Channel 9 of CCTV. This documentary breaks through the simple interpretation of the scientific knowledge of climate change, but analyzes the various effects of climate change on human beings from the macro perspective of human existence and civilization development, contains the abstract content in popular stories, cuts in from human feelings, ends in specific low-carbon action plans, and allows the audience to more realistically understand the relationship between climate, environment and people through the use of comprehensive frameworks. This documentary was very popular during the premiere and the simultaneous video clicks exceeded 3 million.

It is worth noting that the role of the Chinese media in climate change communication and governance is becoming clearer and the strategy is constantly adjusted.

However, with the rise of online media and self-media, the traditional media industry has begun to shrink. External and internal reasons, such as strong "seasonal characteristic" of climate change issues, few hot information usually and higher requirements for professionalism, have led to the loss of climate reporters who have grown up, which has caused the development of Chinese media in the field to slow down a certain extent.

3 Role Changes of NGOs

NGOs are divided into international and local ones. According to the definition of the UN, international NGOs refer to organizations that have not been established under the intergovernmental agreement. Local NGOs are self-organized civil organizations with the goal of promoting public interests such as sustainable development. Due to the charitable, non-governmental, and non-profit characteristics of NGOs, some international NGOs with an international influence were granted a position of "observers" by the UN at an early stage. International NGOs use "conceptual cognitive power" such as information, persuasion and moral pressure to promote the reform of international institutions and governments. Under the demonstration of international NGOs, more and more local NGOs bloom. The talents cultivated by international NGOs have become the founders or backbones of local NGOs, and the right to speak of NGOs as an independent actor has been strengthened in the global governance.

In the Copenhagen Negotiation, local NGOs have a large gap in both professionalism and flexibility with international NGOs, with lower scores in the analysis of interested parties. With the deepening of international climate change negotiations, the role of local NGOs has gradually been recognized by the government and the media from 2009 to 2015. The interaction and cooperation between the Chinese government and NGOs in the field of climate change has become more frequent and regular.

3.1 The Role of NGOs in COP15

Role 1. Negotiation monitor

Promoting the development and implementation of sustainable development goals is the mission of NGOs. Taking climate change as an example, initially, when the global public knew little about climate change, NGOs enlightened the public through various advocacy actions to make them be aware of the dangers of climate change and started from themselves and acted aggressively to fight against climate change. In international climate change negotiations, NGOs serve as an independent third party to oversee the fairness of the negotiation process and to balance the game forces of developed and developing countries.

Table 1 Basic information of three international NGOs that follow the progress of China's climate governance at two levels

Agency	Founding time	Headquarters	Scope of work	Focus of climate
WWF	1961	Switzerland	Protect the diversity and living environment of living creatures in the world	Mitigate and adapt to the impact of climate change on the biodiversity and environment
Oxfam	1942	Oxford, England	Comprehensive development of emergency relief, poverty alleviation, education, health, etc.	Protect the interests of poor countries and people most affected by climate change and emphasize the importance of adaptation
Green peace	1971	Amsterdam, the Netherlands	Environmental protection	Impact of climate change on the environment, and emphasis on emissions reduction

Source: Prepared by the author

The Copenhagen Negotiation attracted the participation of more than 20,000 NGOs. From the perspective of guiding the public opinion, the most influential international NGO in the field is the World Wild Fund for Nature (WWF, founded in 1961, headquartered in Switzerland), Oxfam (founded in 1942, headquartered in Oxford, England and later moved to Nairobi), and Green Peace (founded in 1971, and headquartered in Amsterdam) (see Table 1). Since the launch of global climate governance, the three organizations have participated in the entire process with a goal of safeguarding the common interests of mankind and contributed to the signing of the UNFCCC and the Kyoto Protocol.

The three organizations in climate change negotiations all aim to urge the developed countries to achieve the emissions reduction and adaptation commitments as soon as possible from the perspective of developing countries and to take facilitating the conclusion of fair, just and binding agreement as a phased goal. The commonalities of the three organizations also include establishing branches and sub-branches in Mainland China, arranging special office spaces in Copenhagen, having global media resources and credibility, having international experts with long-term tracking of climate change negotiations, having professional teams to be responsible for handling international media affairs, and having delegates who are invited by governments of various countries to participate in government Delegations.

In international negotiations, international NGOs are natural allies of developing countries. As mentioned earlier, developed countries have dominated the process of global governance for a considerable period of time and occupied the absolute discourse power. Developing countries including China can only follow passively.

To this end, the three organizations take helping developing countries to speak as the main work strategy to balance the discourse hegemony of developed countries.

Taking Lesotho as an example, Lesotho is a small countrylocated in southeastern Africa. The country is completely surrounded by South Africa and has few natural resources. It is one of the least developed countries announced by the UN. Faced with the threat of climate change, Lesotho with high vulnerability is the most direct victim of climate change and has little resilience. Countries such as Lesotho are the most vulnerable in the international game and their voices are most easily overlooked. Limited to human and financial resources, the national Delegation of Lesotho could not organize a press conference during COP15. To help the Lesotho Delegation voice out, Oxfam invited international media agencies including BBC and CNN to make a special interview of Lesotho Prime Minister. The interview and report on the threat of climate change to Lesotho helped the international media to have a new understanding of the fragile forces in the international game while Lesotho won the right to speak. In this way, Oxfam promoted developed countries to fulfill their commitments to ensure the fairness of the negotiations.

From the perspective of work areas and priorities, WWF and Green Peace are international environment-based NGOs, while Oxfam is an international development-oriented NGO based on humanitarian relief and poverty alleviation. The difference in priorities determines that the three organizations are different in views and perspectives but complement with each other. In the second week of the Copenhagen Negotiation, the developed countries refused to fulfill the international commitments and became the biggest obstacle to the negotiations. WWF, Green Peace and Oxfam held a joint press conference in the venue to criticize the behaviors of developed countries from their respective expertise and to prevent the rights and interests of developing countries from being infringed.

Role 2. Negotiation catalyst

As an independent third party, NGOs are invited to participate in the UN climate change negotiations and have an opportunity to observe subtle changes in the negotiating landscape in the process of fulfilling the role of supervisors. If the organization itself is sufficiently professional, it can seize the opportunity in the subtle changes and promote the negotiations to be carried out under the premise of openness and fairness through communication with the media and negotiators.

WWF, Green Peace and Oxfam have more than 20 years of experience in the history of climate change negotiations as well as a clear division of labor based on expertise. Taking Oxfam as an example, more than 80 staff members of its Copenhagen Delegation are assigned to five professional team, namely expert team, media team, activity team alliance team, and Delegation team.

The expert team consists of experts who have followed up the negotiations for more than 10 years and can conduct real-time analysis, interpretation and pre-judgment based on scenario assumptions to provide professional comments based on external needs.

The media team consists of staff members with many years of media experience and rich media resources. They are responsible for docking with the media and expert team, collecting media questions, responding to replies of experts, arranging interviews, coordinating press conferences, etc. Oxfam has a very good reputation and influence in the international media. When the media faces too much professional negotiation contents that cannot be understood or does not know how to follow up reports, they will take the initiative to find Oxfam staff. In response to the inadequate preparation of the Chinese media, Oxfam has specially organized trainings for them.

The members of the activity team are experienced activists good at advocacy and mobilization. This team can design flexible and diverse reality shows, protests, melodrama, etc. according to the progress of the negotiations and Oxfam's position to guide international public opinions through media reports. For example, on the first day of the negotiation, the activity team arranged the representative from Maldives to be loaded in a specially treated water-filled device and holds a sign in his hand "please save us as we are to be flooded", which has become the headline of major international media.

The alliance team is special as its members are not Oxfam's staff. Oxfam's focus is development and has a broad network across the grassroots communities in developing countries. The alliance team opens its quota to Oxfam's partnership network and invites representative partners to COP15 to voice out on behalf of the vulnerable people. Meanwhile, the participation can help the local partners to build up their capacities.

Members of the Delegation team have dual statuses. One is Oxfam's staff or its partner agencies' representatives and the other is members of national government Delegations. Some countries specifically invite representatives of NGOs working in their countries to join the government Delegations to coordinate and balance the standpoint of the Delegations. Government Delegations of the Philippines, the Netherlands, Belgium, Bangladesh and other countries contain representatives of NGOs. Joining the government Delegations means that international NGOs have direct access to the negotiations and more effectively participate in the global climate governance.

Just because of the above-mentioned professional division of labor, international NGOs have been very flexible in the negotiation in Copenhagen, and they are able to discern valuable contents from information obtained through formal and informal channels. The most typical example is the Danish text exposed by *The Guardian*, which is disclosed by international NGOs to *The Guardian* after receiving the original text at the first time to reach the goal of advancing the negotiation process through media.

3.2 Tracking Analysis: Strategy Shift of NGOs at the International Level

1. Change from watchers to collaborators

By sorting out the roles of international NGOs in the Copenhagen Negotiation, it can be found that NGOs play a special role of supervision and catalysis in international negotiations. International NGOs have been following the process of UN climate change negotiations for a long time, and they have relatively mature strategies to play the roles.

From 2009 to 2015, the game relationship between international NGOs and the Chinese government has undergone subtle changes.

In Copenhagen, international NGOs are China's potential game objects. On the one hand, international NGOs see China's high emissions facts; on the other hand, most international NGOs respect the principles of the UNFCCC and China has a certain moral advantage as a developing country. Even in the end, international media leaned to one side, but China has not become the target attacked by international NGOs. For China that is the first high-profile time to participate COP, while international NGOs generally hold a wait-and-see attitude. In the ever-changing international climate negotiation process, international NGOs prefer to judge their attitude to China based on China's follow-up performance. In the follow-up negotiations, the Chinese government's attitude towards NGOs has become more and more open, international NGOs have smooth access to China's information and their understanding and awareness of the China's reality has gradually increased. In particular, WWF, Oxfam, Green Peace, the World Resources Institute, the Energy Foundation and other international NGOs with branch offices in China can observe the actions of the Chinese government both at the international and domestic levels. At the domestic level, the Chinese government's policies and series of actions to address climate change can be observed at the first time; at the international level, China's determination to participate in global climate governance can be felt from contacts with negotiators and various sources of information. Combining with the comprehensive information on both international and domestic aspects, such international NGOs take the first to become the alliance of the China in the climate field. On the basis of the "common but differentiated responsibilities" principle, they work together to urge developed countries to fulfill their emissions reduction responsibilities. When international public opinions violate the principles of the UNFCCC, such international NGOs can provide a strong support from the perspective of climate justice.

As China becomes the world's largest emission country, the voice that China should assume the responsibility of reducing emissions has been an excuse to be used by developed countries to shirk their historical responsibilities in the follow-up negotiations. During the COP17 in Durban in 2011, the international Climate Action Network (CAN) nominated China and India for "the Fossil of the Day", which is a campaign event that CAN has adhered to for 12 years and elects the country with the worst performance in the negotiations of the day. Countries such as Canada, the

United States, and Japan, which are trying to obstruct the negotiation process, have been on the list, and negotiators of these countries have also been forced to receive the "award". The award is set to be directly related to the negotiation process and is the spotlight moment of the international media. China and India were nominated because some representatives of CAN proposed although China and India belong to developing countries, considering the economic strength and international influence of the two countries, if the two countries can make more commitments in emissions reduction, it will have a direct push on the negotiation process. According to the voting rules, if the NGOs of the nominating countries unanimously oppose it, renomination should be made. International NGOs such as Green Peace, WWF and Oxfam quickly talked with representatives of China's local NGOs and helped them to prepare the reasons for response. At the voting stage, representatives of Chinese NGOs did not pass the nomination. At the same time, Indian NGOs also voted against it, avoiding misleading international public opinions.

Before the 2015 climate change conference in Paris, the voice that China should take more responsibilities began to be hyped. Oxfam published a report titled *Extreme Equality in Carbon Emissions*, which further highlighted the issue of climate injustice that "the richest 10% of the population produces about half of the world's carbon emissions, while the 3.5 billion poorest people only produce 10% of carbon emissions, but the latter's life is threatened by superstorms, droughts and other extreme weather events related to climate change." The report pointed out that despite the rapid growth of carbon emissions in developing countries, it should be noted that most of the commodities produced by developing countries are provided for the people of developed countries to consume and use, which means that the carbon consumption of living in most developing countries is much lower than that of developed countries. The content of this report was widely reported by the Reuters and other media, which not only clarified the fact that developed countries have shifted emissions, but also once again responded to the problem that developed countries shrink their responsibilities.

2. Helping local NGOs grow

Compared with the performance of international NGOs in Copenhagen, although China's local NGOs such as China Youth Climate Action Network (CYCAN), Shanshui Nature Conservation Center, and Beijing Global Village Environmental Education Center are present, they are similar to the Chinese media to participate in professional negotiations for the first time and can only passively follow the negotiation process and have few ability to participate in depth.

In the follow-up negotiations, international NGOs helped local NGOs improve their understanding of climate governance issues through talent development and delivery, capacity building, project cooperation, joint advocacy and other ways.

Joint action is an advocacy strategy widely adopted by international NGOs. When participating in the Copenhagen Negotiation, China's local NGOs were weak and inexperienced, and they did not have a clear understanding of the role of joint action.

Based on observations and learning in Copenhagen, international NGOs and local NGOs made a meaningful attempt in the Tianjin conference 10 months later.

The UN climate change conference in Tianjin was a phased negotiation in 2010. The task was to prepare for the COP16 in Cancun at the end of the year. Compared with a few months ago, the number of local NGOs participating in the negotiations increased to more than 60. In order to play a better role in the negotiations, WWF, Green Peace and Oxfam, together with the local NGOs discussed cooperation strategies two months ahead and set up a special preparatory committee. The preparatory committee regularly coordinated the participating organizations to hold a joint meeting to integrate the latest information of all the parties and discuss the participation strategy. The preparatory committee also drafted a common position for publication in the negotiations. After careful planning, more than 20 series of activities were launched in Tianjin. The topics covered the green innovation of Chinese enterprises, the impacts of climate change on China, and the direction of China's emissions reduction, etc.

For better communication and interaction, China Climate Change Action Network (CCAN) began to play a coordinating role with the help of international NGOs. To enhance professionalism, CCAN has invited experienced international organizations such as Oxfam, Green Peace and WWF to provide advices as observers. In 2011, China Climate Policy Team (CPG), which is dedicated to improving the participation of local NGOs in international negotiations and the ability of advocating domestic policies, was established. CPG more focuses on the negotiation and policy level than CCAN, and its members are more stable and have played a positive role in the follow-up negotiations. In addition, local NGOs such as CYCAN have also grown up and further enhanced their capabilities in the cooperation with international NGOs.

The reason why international NGOs are respected in climate change negotiations is closely related to their long-term accumulated research experience, strength and professionalism. After the COP15 in Copenhagen, with the guidance and assistance of international partners, China's local NGOs also began to consciously accumulate research experience in this area. Greenovation Hub, which focuses on climate policy research, is one of the typical representatives. After its establishment in 2012, it has published many influential research reports in the climate and energy fields and won recognition from international counterparts.

3.3 Tracking Analysis: Strategy Shift of NGOs at the Domestic Level

1. Refining game objects and exploring cooperation space

The main task of international NGOs at the international level is to monitor the negotiation process based on state actors and ensure the fairness of negotiations. In China, the main role of international NGOs is to ensure the domestic climate policies to be forward-looking and effective. After the Copenhagen negotiation, international NGOs with offices in China have increased their focus on the climate sector and opened up more space for cooperation relying on their respective expertise to help the China government fulfill its international commitments.

Considering China's domestic policy environment, different international NGOs are actively exploring their working space.

Taking Green Peace as an example, Green Peace was solidified as a radical environmental organization due to the image that its staff members drove a boat to intercept a whale catcher. In order to carry out work in China, Green Peace has adjusted its strategy based on the analysis of domestic game objects to strive for win-sets and ensure goals to be achieved. Under the policy of the central government to control fossil fuels such as coal, Green Peace monitored the implementation of enterprises using its skilled in-depth investigations. In July 2013, Green Peace issued a report accusing China's largest coal company Shenhua Group of over-exploiting groundwater and polluting local ecology in Ordos, Inner Mongolia. After a 255-day campaign, Shenhua promised to gradually stop the extraction of grassland groundwater (Feng, 2014). From April to October 2014, Green Peace's investigators conducted field investigations on Datang's coal-to-gas demonstration project in Hexigten Banner that was taken as the first to be put into operation, carried out sampling and testing of sewage and sediments of the project, analyzed the monitoring data of smoke emissions, and interviewed the herdsmen around the project site of their living conditions. The on-site investigations found that the demonstration project had serious environmental pollution problems. On August 7, 2014, Green Peace released a report stating that after three years of investigations, it found that a number of enterprises in the mining area of Muli Coalfield in Qinghai Province carried out illegal open-pit coal mining and illegal projects in the water source of the Yellow River in the Qilian Mountains at an altitude of over 4,000 meters. Green Peace was under pressure from the local governments and enterprises in the process of carrying out these projects. The interesting point in the China context is that the local governments and enterprises are also the game objects of the central government in domestic climate governance.

Therefore, these pressures have not really affected the advancement of overall work of this agency in China.

WWF's domestic efforts to address climate change are dominated by research and public advocacy, and its low-carbon city project focuses on China's transformation of transportation modes in the process of industrialization and urbanization. By supporting partners to write research reports and advocating the consideration of improving energy efficiency, low-carbon transportation, improving water quality, etc. in the new urban design, it has provided success stories for China's low-carbon transition and set a new example for global sustainable cities. At the same time, WWF's global campaign "Earth Hour" also was introduced to China in 2009 and made an growing influence.

Oxfam's positioning in addressing climate change is to emphasize the relationship between climate change and poverty. Oxfam has been working in Mainland China for more than 20 years, and its long-term experience at the grassroots level has enabled it to understand China's situation affected by climate change at the first time. Based on the project's advantages, Oxfam set up a dedicated climate change and poverty project team and worked with research institutions to carry out research on climate adaption, resilience, climate finance, climate change and food security. Oxfam comprehensively promote the systematic development plan on Low-carbon Adaptation and Poverty Alleviation (LAPA). The first pilot of LAPA is in a rural village in Shaanxi province. After two years of grassroots work, it has successfully advocated the local government to integrate the perspective of climate change in the process of designing local development plans to help poor rural areas shift from passive adaptation to active adaptation.

2. Building work network

Network building and partnerships are effective strategies for NGOs to increase their influence. After the COP15 in Copenhagen, international NGOs simultaneously expanded their partnerships in the climate field in China, and local NGOs also increased their efforts in joint action and grew rapidly.

Local NGOs that are active in the field of climate include Greenovation Hub, Global Environment Institute, Shanshui Nature Conservation Center, Friends of Nature, Beijing Global Village Environmental Education Center, Green Earth, the Institute of Public and Environmental Affairs (IPE), CCAN and CYCAN. From 2009 to 2015, local NGOs have carried out a lot of active cooperation and co-built their influence through mutual resource advantages. The Green Commuting Fund of CANGO organized the "cool China – national low carbon action plan" in various provinces and cities in 2011 to promote low-carbon travel. Nearly 40 NGOs launched the "climate citizen transcendence action plan" in 2013 to promote multi-stakeholder joint response to climate change. Since 2014, NGOs have also established a China climate policy team to coordinate the government, think tanks and NGOs to carry out cross-border dialogues on climate change issues and participate in the discussion and revision of legislation on addressing climate change. In 2014, after IPCC released the fifth assessment report, seven NGOs including Greenovation Hub, Shanshui Nature Conservation Center, and Global Environment Institute, made a joint response to encourage the Chinese government to pay attention to the multiple challenges brought about by addressing climate change and take measures at multiple levels to address climate change.

3. Promoting mobilization at the grassroots level

NGOs have a natural advantage in community work. According to the 2012 survey of China4C, the Chinese public has a high awareness of climate change and a high degree of support for climate policies, but the willingness to take action is not particularly strong. So, local NGOs focused on popularizing knowledge on climate change to the front line and developed some operative tools for energy conservation and emissions reduction to promote effective climate actions.

Take the low-carbon family project of Friends of Nature as an example. Friends of Nature has promoted pilots in more than 200 families in Beijing since 2010 and built the most economical electricity consumption mode based on the electricity consumption of these families. Based on this pilot work, Friends of Nature developed a user-friendly toolkit which is available to more families.

In order to better monitor the greenhouse gas emissions of companies and respond to the smog problem of public concern, IPE has developed a mobile application software as pollution map that allows people to view the air quality rankings of cities across the whole China, understand the root causes of industrial pollution, and share pollution intelligence with friends and protect the environment together. The map can indicate the name of companies that exceed the discharge standards and mark the relevant indicators of harmful gas emissions of companies so that the public can see at a glance whether companies exceed the standards or not. The public can share information to the social platforms at any time so that companies can accept public supervision and the public can find specific action points.

At the provincial level, there also are a number of grassroots organizations that are interested in climate change. Shaanxi Volunteer Mothers Association for Environmental Protection insists on the guidance of garbage sorting and recycling in rural areas; Shaanxi Province Science and Technology Service Center for Rural Women, established by several professors of Northwest A&F University, brings the latest technologies to rural families, studies insect variation under climate change, helps farmers install solar insecticidal lamps, and writes observations at work into proposals for the central government to attract more attention on local level at a more macro level.

4 Main Conclusions

By reviewing the Copenhagen Negotiation, it can be seen that the Chinese government, media and NGOs are generally inadequately prepared, exposing some problems i.e. the rigidity of mechanisms, inadequate plans and weak professionalism, etc. From the perspective of two-level game, this is the spillover of that the Chinese government did not form a consensus on the importance and strategic value of climate change.

As the main actor of negotiations, the Chinese government has put forward the target of voluntary emissions reduction before the negotiations and also sent the biggest Delegation in history. However, as the world's largest emission country and the second largest economy, the national plan came up by China still has a certain gap with the expectations of the international community. It was the first high-profile appearance of the Chinese Delegation in the Copenhagen Negotiation and at that moment, the Delegation lacked the negotiation experience and game ability. Therefore, in the second week, in the face of more than expected negotiation trends, the Chinese government Delegation was very passive. As the main information resource, the Chinese government made a progress by trying to hold press conferences to share information However, due to the administrative system and inertial thinking, the distribution channels and contents still lagged behind, affecting the timely release of information. In the Copenhagen Negotiation, the government also had the role of an administrator. Global governance emphasizes the diversity of subjects. The foundation of the governance process is not control, but coordination. It is a continuous interaction and emphasizes the participation of all interested parties in the process. To participate in global governance, China must change the traditional way of thinking.

The Chinese media also temporarily rushed to Copenhagen in the absence of preparation. As the climate communicator, the Chinese media is able to play a role on the international level. But as the negotiating booster, the Chinese media is far behind the international counterparts. In the last two days of the negotiations, when China was accused by the international media, the Chinese media that was accustomed to being managed by the government and lacked an independent speculative ability did not even have a basic debate. Because the role was not adjusted in time, the Chinese media did not exert an influence on the international stage in COP15.

International NGOs have actively participated in the early stage of global climate governance to accumulate a certain amount of experience and have relatively clear positioning and strategies and played the role as a monitor and catalyst in COP15. In contrast, China's local NGOs lacked experience in global governance and had unclear positioning and limited capabilities at that moment.

It can be seen that the Copenhagen Negotiation was a baptism for the Chinese government, media and NGOs and was the real reflect of the domestic governance style at the international level. Because of the lessons and serious reflections, the three key stakeholders have made rapid adjustments to the strategies for climate change communication and governance.

First, the government is aware of the difference between the international and domestic arenas. It has begun to explore more flexible negotiation strategies and information dissemination methods for the international stage, and the interaction with the media and NGOs has become more frequent. At the same time, due to the pressure of international public opinions and the need for its own development, the government has raised addressing climate change to the level of national strategy. In order to better enhance the confidence and determination of all sectors, the government has consciously transmitted the pressures in international negotiations to the domestic level, promoted all stakeholders to implement the emissions reduction targets, and actively mobilized the public to participate. Because of the updating understanding of the role of NGOs at the international level, the government is relatively open to NGOs in the climate field at home, the communication and cooperation are more frequent than in other areas. To some extent, the climate field is an experiment pilot for a new governance style in China. The government's domestic response to climate change has also been passed on to the international community to help China to win more supports in international negotiations so that China can play a more constructive role.

Due to the study during the Copenhagen Negotiation and the subsequent changes in the government attitudes, the Chinese media is increasingly aware of the role and seizes the space released by the government to report the climate change issues and the negotiation process from a more comprehensive perspective. At the international level, the Chinese media has consciously helped the international community to understand the development reality of China and spoken for developing countries. At the same time, Chinese media has also actively transmitted the international pressures to the country. Domestically, climate change has become a key area of media coverage because of the government's high priority. In the reports, the reporting's perspective is more comprehensive, and the discourse framework is more diverse. Although there is still room for further improvement in the role of negotiating boosters, the Chinese media has learnt to strategically disseminate climate change information at the two levels to produce an interactive impact.

The growth environment of NGOs in China is not smoothy and the foundation is weak. However, because the government realized NGO's particular role in the global governance, coupled with the demonstration and experience sharing of international NGOs, local NGOs in the field have grown rapidly. Many international NGOs which have branches in Chinahave turned to be collaborators and allies. The local NGOs have turned to joint action from individual action, actively accumulated action and research experience, and been gradually able to assume the role of monitors as international NGOs. Domestically, NGOs' division of labor is more meticulous and they are more positive in cooperation and good at down-to-earth in implementation. The results of these advancements have been brought to the international stage and won the respect of international counterparts.

Based on the above analysis, the Copenhagen Negotiation played the vital role for China to learn important experience in dealing with the future. Since then, the international rules and regulations on climate have been rapidly localized and the subsequent international climate change negotiations have further pushed the Chinese government, media and NGOs to adjust their strategies at the international and domestic levels in a timely manner.

On December 13, 2015, 196 contracting states, including China, reached the Paris Agreement to make institutional arrangements for the global response to climate change after 2020. The Paris Agreement is the result of the joint efforts of all parties and China played a key role in the process.

In the six years from 2009 to 2015, China has grown from a follower to a leader at the fastest rate. Through the analysis of the two-level and multi-dimensional research framework, we can see that the three key stakeholders have worked together to embark onto an efficient climate governance road with Chinese characteristics from single-layer to double-layer cognitions, from individual to cooperative attitudes, and from management to governance concepts.

As the world's second largest economy, China's strength has been difficult to maintain a low profile on the international stage. In the context of global governance, China can play a more constructive role in the multilateral process. "The world is shifting to a new order, China has a place in this new order, and new rules have begun."[10]

---

[10]Comments made by Jonathan Holslag, Chairman of the Brussels Institute of Contemporary China Studies, in an interview with the *21st Century Business Herald*, searched on http://news.Lnd.com.cn/htm/2009-12/12/content_962061.htm.

Chapter 6
China's Climate Change Communication and Governance in the Post-Paris Era

1 China's Challenges

Addressing climate change requires both global cooperation and local action. The goal of climate change communication is to contribute to the global climate governance. The particularity of climate change issues determines the necessity of two-level research. Climate change communication aims to help China play a more constructive role in international negotiations, so as to promote more effective global cooperation. In addition, climate change communication aims to make more domestic stakeholders actively participate in climate change response. From the COP15 in 2009 to COP21 in 2015, the Chinese government, media and NGOs have made some progress, but still face challenges in the post-Paris era.

1.1 Continuous Enhancement of Right of Speaking

With the rising of international influence, China has become a key factor affecting international climate rules. As one of the great powers in the world, China should have good spiritual outlook, political reputation and shoulder the international obligations and responsibilities endowed by the times. The growth of a great power usually goes through three stages, which are preparation, growth and stability. As a great power in the growth stage, China needs to enhance its own strength and establish a "responsible great power" image through concrete actions that the international community is convinced about.

China's image has been marked with various negative labels in the past few years. Although China has made many positive contributions to global climate governance, and is communicating through various channels, China still needs to enhance its right of speaking and influence in climate governance as long as the game of emission rights continues.

Before the COP21 in Paris at the end of 2015, the Chinese government signed bilateral and multilateral climate cooperation agreements on the basis of the China-U.S. joint statement and China-France joint statement, and contributed 20 billion yuan for South-South Cooperation on Climate Change, demonstrating the ambition to promote climate governance and winning the initiative to participate in the negotiations. However, on December 10 (the second week of negotiations), the *Guardian* published a signed article, stating that 195 countries including the U.S., the European Union and Alliance of Small Island States, constituted the High-ambition Coalition, to promote the conclusion of Paris Agreement. *The Guardian* emphasized that this Coalition does not include China and India. According to the provisions of the UNFCCC, developed countries should assume the historical responsibilities of reducing emissions and provide developing countries with financial and technical support for climate change response. According to the Copenhagen Accord signed in 2009, developed countries should provide the fund of US$100 billion annually to support developing countries in addressing climate change by 2020. However, before the negotiations in Paris, the support fund was far from being put on the table, and only 16% of the adaptation fund urgently needed by developing countries was allocated. Therefore, the issue of funding became the focus of negotiations in Paris. The U.S. and the European Union were criticized by the international media in the first week because they did not honor their commitments on funding. Just in this context, *The Guardian* published an article, which transferred the pressure to two developing countries China and India by taking advantage of the audience's reading habits and psychological expectations. The Chinese government has established an international public opinion monitoring mechanism based on the lessons learned over the past few years in order to respond in a timely manner. After noticing the report of *The Guardian*, the Chinese government invited the Xinhua News Agency to write relevant reports based on objective facts. However, due to the declining of traditional media in recent years, difficulty of tracking and reporting climate change issues, and lack of understanding about the history of Negotiation, the Xinhua News Agency did not give any response until the next morning, thus losing the best timing of news. In addition, the tone of their reporting mainly targeted at domestic audience rather than the international community. As a result, public opinions continued to be unfavorable to China. After analyzing the potential risks, the Chinese government delegation temporarily organized a press conference, inviting international media to attend, which avoided more serious consequences to China.

1.2 International Expectations and China's Action

With the improvement of national strength, China has become an indispensable key role in the discussion of various international regulations. Attaching more importance to climate change, China has played a more proactive role in the international arena. China is one of the world's largest emitters and one of the fastest-growing developing countries, so the international community raises higher and higher expectations for

China. For example, during the climate negotiations in Paris, I received the interviews by international media such as Reuters. I was repeatedly asked about China's responsibilities: "besides previous commitments, what else can China as the world's largest carbon emitter do?"; "What more can China contribute to the Paris Negotiation?" It can be seen that although China has made many voluntary contributions beyond historical responsibility, the international community still has many expectations for China.

As a large developing country, China is undergoing industrialization and rapid infrastructure construction, has strong energy demand, and encounters certain technical constraints in improvement of energy efficiency, which determines China's greenhouse gas emissions will still increase in the near future. Decided by the objective rules of economic and social development, it is the real difficulty faced by China. China cannot meet the expectations of the international community in one step. In this case, China needs more appropriate expression to help the world understand China properly and resolve the pressure of international public opinion. Based on the lessons accumulated in the past six years, all parties of China have been aware of this challenge, but still need improvement in response strategy. During the climate negotiations in Paris, a side event on "climate change and rural development" was held in the China Pavilion, to show the international community the reality of China's rural poverty and the urgency of adapting to climate change. This is the first meeting of rural theme since the establishment of the China Pavilion by the Chinese government delegation in 2011, which makes the world have a better understanding of China. Although several research reports on China's climate change and poverty issues were issued, and guests from Ethiopia, India, Uganda and other countries delivered speeches and shared the cases, that meeting, as ever-changing negotiations entered a crucial stage, could not attract the attention of the international media and failed to achieve effective communication effects.

1.3 The Largest Win-Set not yet Formed

The two-level game theory integrates international and domestic politics, emphasizing the game of decision makers at both the international and domestic levels. The ultimate goal of signing any international climate agreement is to urge each country to replace the traditional development model based on high emission with a sustainable low-carbon development model. Only by gaining the largest win-set and winning most stakeholders' support can China really change to the sustainable low-carbon development and contribute to the international climate governance.

Through the two-level tracking research of strategy shift of the three major stakeholders, we find that the government, media and NGOs have become more and more interactive in addressing climate change. However, other stakeholders, especially the public and enterprises still need to be further mobilized, and the largest win-set has not been formed yet in China around 2015. Survey results show that the Chinese people have a strong awareness of climate change, but they still need to be further mobilized to take action.

2 China's Strategy

Based on the two-level game framework, international and domestic levels are mutually influential. To tackle the above three challenges, we cannot separate the international and domestic levels. Climate change communication and governance strategies for the post-Paris era can be analyzed in the "dual-layer, multi-dimensional" research space constructed in this book.

2.1 Telling Stories of "Real China" from Domestic to International

From the perspective of international relations, the government, media and NGOs are all transnational communication actors, which refer to "the actors that communicate between internal and external political environments for specific political goals, so as to achieve the cross-border flow of values, norms, new concepts, policies, regulations, goods and services" (Su, 2003).

The emergence of transnational communication actors subverts the division of international and domestic levels in the traditional sense. Transnational communication actors can shuttle freely within the domestic political and economic system, directly introduce international rules to domestic decision-makers, or raise domestic issues to the international level so as to challenge the closed domestic decision-making process. Without transnational communication actors, the international system and domestic politics cannot be linked.

As a country directly affected by climate change, China has made tremendous efforts to address climate change, but it has not been fully known by the international community despite efforts of six years from 2009 to 2015. China habitually shows off its achievements on the international stage, being unwilling to mention the practical challenges such as uneven development and rural poverty, which makes many international friends who only know Beijing, Shanghai and Guangzhou deepen the wrong impression that China has become a developed country, also bringing more international pressure to domestic development.

To solve this problem in the post-Paris era, China must give top priority to revealing the "real situation" to the world.

First, Chinese media should improve the reporting strategy. In order to make China better participate in global climate governance, the media should further improve the international climate change communication skill to help the international community correctly understand China's national conditions and contributions to climate change response under limited conditions. Chinese media should avoid the order-based reporting and empty and boring didactic publicity, and tell vivid stories to foreign audiences, especially actual national conditions of China.

Second, we should invite international media to report "real China". The international media is a definitive stakeholder of China in climate change communication, but there is uncertainty in correlation. If China's actual national conditions and actual contribution to climate governance are introduced to the international media in a timely manner, it can enhance the positive correlation between international media and China.

In cooperation with international media, Chinese governments at all levels need innovation and more courage and confidence. In June 2013, the Chinese government launched the first "National Low Carbon Day", organizing a number of interesting activities to publicize the "green and low-carbon" concept, which is also unprecedented in the world. However, no international media was invited to attend the opening ceremony of this national campaign. Another example is that Chinese government invited international media to interview the low-carbon demonstration cities and experience China's rapid development. However, as they visit more and more affluent cities, the international media raises higher requirements for China in reducing emissions. The international media hopes to know about real China, but their interview requests are usually rejected for a variety of reasons. Local governments frankly admit that they fear the international media may discover the problems of rural areas such as poverty, backwardness, and local conflicts. Finally, we should attach importance to the special role of NGOs in conveying the "real China". The authority of international NGOs has been recognized by the international community for their expertise in climate change and high-quality analysis. International NGOs engaged on climate change work in China understand China's actual situation and challenges after the Copenhagen Negotiation, Chinese local NGOs have consciously enhanced their abilities, and made valuable researches from the perspective of the developing countries, having made constructive contributions to changing the "one-sided reporting" situation on the international stage. A number of local NGOs that have long been rooted in the grassroots units better understand China's national conditions, having collected a lot of first-hand information. The Chinese government can cooperate with such grassroots NGOs to better tell Chinese stories and help the international community understand the real China. Only by shaping the modern partnership between government and NGOs can China get more ideal outcomes of governance.

2.2 *"Pressure Transmission" from International to Domestic*

The objective law of economic and social development determines that China's greenhouse gas emissions will continue to increase in the near future, China cannot assume the responsibility beyond its scope of capacity, and China's goals of development may conflict with global emissions reduction tasks at any time. Properly making the domestic stakeholders aware of China's pressures from the international community may win the maximum support on emissions reduction at the domestic level. At the same time, China may strive for more understanding at the international level through effective communication of the real stories. Only in this way can China turn challenges into opportunities.

First, the media should further enhance the professional capabilities in Pressure Transmission.

The pressure from the international community should be conveyed properly. Excessive stressing of international pressure will make the people lose confidence in the Chinese government and repel China's participation in global governance; intentionally downplaying pressure will make the public unable to understand the challenges the Chinese government faces in a timely manner so that they have no ambition to achieve low carbon development and emissions reduction.

After six years of accumulation, Chinese media's climate change reporting has become more objective, comprehensive and diverse, but there still exist several problems. First, report density is not enough. Reporters usually focus on reporting or making comments on extreme weather events or climate change conferences, lacking the follow-up reporting, with serious content homogeneity. Second, the emphasis on climate change reporting is still insufficient. No media has fixed layout for climate change reporting, which to some extent affects the reporter's enthusiasm. In addition, compared to professional international media, Chinese media still lacks global vision and scientific cognition, which may affect the extent of pressure transmission and destroy the expected effect.

In order to accurately perceive and transmit the international pressure, the media should have the awareness and ability to set agendas, follow up the issues of negotiations for a long time, collect and systematically analyze international public opinions. At the same time, the media should make in-depth reporting of domestic climate policies and practices from various angles. Only by making efforts at both the international and domestic levels can the media achieve the in-depth, accurate and powerful reporting and transmission of climate change information.

In the process of domesticating international climate regulations, the media should make the domestic people realize the world's common cognition and determination to address climate change, so as to raise the Chinese public awareness about the severity and urgency of climate change. Meanwhile, when climate change governance becomes difficult or stagnant, the media should strengthen the reporting of relevant government policies to make the public aware of the effective governance measures and firm the confidence in jointly addressing climate change.

Second, NGOs can also play a positive role in pressure transmission.

International NGOs carrying out climate change work in China not only understand the situation of international negotiations, but also the actual conditions of China, thus having unique advantages in transmitting the pressure of international regulation. When China faces the international public opinion pressure, sensitive international media will usually raise interview requests. In this case, international NGOs in China, with the flexible internal coordination mechanism, will quickly respond and accept the interviews first, quickly launching transnational consultations. The staff of relevant countries will quickly sort out and make clear the causes, responsible parties, principles, and stances by mail or telephone. Chinese staff in the international organizations play a key role in ensuring the information related to China is actual. Therefore, the Chinese government should strive for the understanding and support of these international NGOs, maintain regular communication with the Chinese staff working for them, and ensure they fully understand the standpoint of the Chinese government so as to avoid unnecessary misunderstandings. International NGOs can play a buffering role and resolve some pressure, so that the Chinese government can have more time to go through the procedure. In addition, NGOs are naturally linked with the media, and international NGOs have a relatively precise understanding of the international media agenda and discourse framework, which to a certain extent can help the Chinese government respond more appropriately. The local NGOs of policy research growing over the past few years have deepened the cooperation with international media, gradually becoming the information source of international media. Thus, striving for their understanding and support is also important.

2.3 Strategy of Cross-Level "Collaborative Governance"

The two-level game theory emphasizes the interaction at both the international and domestic levels. The premise of full interaction is the joint participation, consultation and cooperation of all stakeholders on the basis of consensus. Climate change is a complex, cross-disciplinary public challenge. Different stakeholders should work together to form the largest win-set and jointly address climate change.

After the Paris Negotiation in 2015, the international community has had higher expectations for China in global change governance. Different stakeholders of China should strengthen consultation and interaction to address climate change in a more innovative way. Correspondingly, climate change communication will be advanced with more solid contents and sufficient motivation. In this way, China will have a larger win-set. If so, China will have more initiative to participate in and play a more constructive role in global climate governance.

1. **Current Stakeholder Relations at the Domestic Level**

Ideally, climate change communication is a dynamic feedback system involving all stakeholders:

The government informs all stakeholders of climate change policies, accepting public feedback and supervision from all parties;

Scientists study climate change issues, spread the knowledge for different stakeholders, and adjust research methods according to the feedback from various stakeholders;

Fig. 1 Dynamic feedback relationship of stakeholders (2015). *Source* Prepared by the author

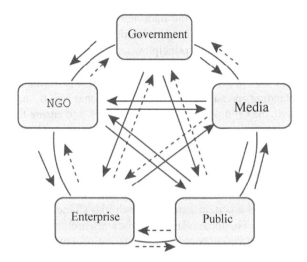

NGOs participate in and monitor the government policy implementation, promote emissions reduction, cooperate with the media to mobilize the public, and feed public opinions and comments back to governments and enterprises;

The media raises the policy awareness of stakeholders, collects public opinions on climate science and policy, and feeds back to governments and scientists, monitoring the actions of various stakeholders;

The public interacts with various organizations on the basis of various climate change information and knowledge to jointly address climate change;

Enterprises consciously implement the government's emissions reduction requirements, make products on the basis of low-carbon development and environmental protection, and urge consumers to practice low-carbon development concept, providing best practice cases for the government and media.

Figure 1 shows the dynamic feedback relationship of stakeholders in climate change communication and governance in 2015, where the solid line represents the ideal link, and the dotted line means the relatively weak link.

Through the analysis of the three major stakeholders above, we can find that the interaction between the NGOs and the media, between the NGOs and the public and between the media and the public in climate change communication has become increasingly regular, but the media monitoring the implementation of government climate policy still remains on the surface, and the mechanism of information feedback to the government needs to be established; after the Copenhagen Negotiation, the Chinese government has tried cooperation with the NGOs and the shift from management to governance, providing NGOs more opportunities to participate in formulation of climate policies.

In addition, other groups of relations are mostly one-way, and the awareness of collaborative governance, and mechanism construction need to be enhanced: The government sets emissions reduction targets for enterprises, while enterprises have no enough channels of feedback to the government; the government formulates low-carbon development policy for the public, while the public needs to really participate in the formulation of such policy; enterprises publicize their energy saving and emissions reduction results through the media, while the media still needs to strengthen monitoring of enterprises; NGOs monitor the enterprises' performance of energy saving and emissions reduction commitments, while the interaction between enterprises and NGOs is not enough. The mechanism of two-way interaction between the enterprises and the public has not yet been established, and commercial relations still dominate.

It is worth noting that Fig. 1 shows the absence of scientists playing a key role in climate change communication. In China, scientists usually play the back-office support role in response to climate change, acting as think tanks for the government, providing information to the media, and cooperating with the NGOs. Although as definitive stakeholders, scientists still need to strengthen the initiative in climate change communication and the two-way interaction with other five kinds of stakeholders.

On the basis of identifying the weak links, we should make the appropriate design to operate the six-in-one dynamic feedback system and mobilize the maximum enthusiasm of all stakeholders to achieve collaborative governance and largest win-set. All stakeholders should promote collaborative governance and more climate change actions at the domestic level, and leverage China to push the global climate governance ahead.

2. Next-step Strategy

According to the above analysis of Fig. 1, we can mainly find the following four problems of domestic climate change response and communication in the post-Paris era: First, the mechanism of stakeholders' feedback and supervision over the government climate policy needs to be improved; second, scientists are absent from climate change communication; third, the enterprises have insufficient enthusiasm to participate in climate change communication and governance; finally, there is a gap between the public's perception and action on climate change.

First, China is continuing the reform and opening up, and gradually building relevant feedback channels. Global climate change is a system involving multiple fields, having the characteristics of large scale, long time span, complex nature, and unbalanced responsibility and impact, etc. Taking the initiative to understand the public psychology appeals through the public climate change cognition survey can enhance the pertinence of relevant policy implementation and social action promotion and solve the problem of insufficient information feedback in the dynamic interaction process to a certain extent.

Second, scientists knowing the scientific principle of climate change can play their special advantages and spread climate change knowledge through different platforms. After the release of the IPCC report in 2014, the National Expert Committee on Climate Change cooperated with the IPCC, introducing the latest findings of climate change science in the campus, which received good feedback.

Third, enterprises are the mainforce for energy saving and emissions reduction. Moreover, they have direct economic exchanges with the public. If enterprise managers truly attach importance to climate change at the strategic level, and also play a special role in climate change communication, the all-win effect can be achieved.

3 Case Study

Wal-Mart was ever criticized by environmental organizations as "a destroyer of the environment". However, subsequently, Wal-Mart kicked off an energy revolution around the world, which not only reversed its brand image, but also changed the consumption habit of consumers. In 2005, Wal-Mart decided to change its image when realizing the energy-saving effect of LED lamps. However, at that time, Americans had used incandescent lamps for 130 years, and LED lamps were expensive than incandescent lamps by eight times, so the Americans were very reluctant to accept LED lamps. Wal-Mart planned to sell 100 million LED lamps by 2008, i.e., doubling the sales of LED lamps in the U.S., which would save the electricity consumption of US$3 billion and reduce construction of a power station. In the face of enormous resistance from traditional thinking and consumption habits, Wal-Mart invited LED manufacturers, scholars, environmentalists, and government officials to gather together to discuss the countermeasures in October 2005. Manufacturers improved the manufacturing process of LED lamps under the guidance of experts. Even so, Wal-Mart sold only 40 million LED lamps in 2005. Afterwards, Wal-Mart began to improve the supply chain, which promoted a new round of price cut and technological transformation. Wal-Mart also took the initiative to ask the environmental organizations to supervise it, expressing the determination to become an environmental leader. In order to make LED lamps conspicuous, Wal-Mart changed the positions of LED lamps on the shelves, further attracting consumers' attention through physical display and banner advertising. Then, Wal-Mart launched a new round of campaign. The producer of former U.S. Vice President Al Gore's documentary "An Inconvenient Truth" helped Wal-Mart design a web page to count the sales of LED lamps in the U.S., making a real-time dynamic map through data visualization technology, through which consumers could visually see the sale situation of LED lamps. As a result, more and more Americans joined the energy-saving campaign. At the same time, Google and Yahoo also helped Wal-Mart promote LED lamps. Americans gradually were accustomed to use of LED lamps. In 2013, Wal-Mart announced to put into operation the first 100% LED-illuminated store in South Euclid, Ohio. LED lamps had the service life more than 25 times that of incandescent lamps, could reduce the energy consumption by 80%, and would generate much less heat than incandescent lamps and compact fluorescent lamps, thus reducing cooling costs. This means that the new Wal-Mart stores would bring energy-saving benefits in the next few decades.

The Wal-Mart case shows that a truly socially responsible company can gain reputation and benefits while promoting public actions and meeting the demands of different stakeholders. In this regard, Chinese companies have made some progress.

China's largest corporate delegation in history consisting of 90 Chinese entrepreneurs from five major business associations and 17 industries attended the COP21 in 2015 in Paris. Mr. Shi Wang, former Chairman of Vanke, a real estate company in China, had been paying attention to climate issues since he participated in the Copenhagen Negotiation, and Vanke had joined WWF's "green emissions reduction pioneer" team, and would track the green supply chain and disclose green building technology and related patents.

On the first day of the Paris Negotiation, 20 countries including China, Brazil, Canada, the U.S., Germany and France jointly issued the "Innovation Mission" initiative, promising to double the government or government-led clean energy R&D investment in the next five years, increase public and private investment in global clean energy innovation and accelerate global clean energy transformation.

At the same time, Breakthrough Energy Alliance was set up, which was witnessed by 28 private sector leaders from 10 countries, including Yun Ma, Chairman of Alibaba Group, Nanpeng Shen, Founding Management Partner of Sequoia China, Shiyi Pan, Chairman of SOHO China, promising to support the commercial application of climate-friendly energy technology. The high-profile appearance of Chinese companies on the international climate stage made the world see China's ambition to address climate change and participate in global climate governance.

Fourth, mobilizing public action is the key to climate change communication. Survey results show that Chinese people generally have strong awareness of climate change, but few of them truly take action.

China4C's national survey in 2017 shows that television is still the main channel for public to get climate change information. Taking into account the seriousness and untouchability of climate change, we can use the theory of entertainment education for the development communication and spread the low-carbon knowledge and successful community practices in an entertaining way, so as to make more people willing to start low-carbon life. In this respect, Europe and the U.S. have made many explorations in the ways of comedy, historical programs, meteorological programs, talk shows, and children's programs. The most successful case is the Hollywood film "The Day After Tomorrow", which made the world aware of serious consequences of climate change. China can also design a TV show themed by low-carbon life or invite a professional team to produce a series of low-carbon life comedy. In this way, more people will better understand the definition of "low-carbon" and "low-carbon life", and the benefits of low-carbon life for the country and the individuals, so that they will take action initiatively.

In addition, we should also pay attention to segmentation research. Having different conditions, different regions have different vulnerability, responsiveness and resilience to climate change, so the people there have different climate change perceptions. The differences in gender, age, and ethnicity make the cognitive differences more complicated. Segmentation research, which means focus on targeted group, is necessary to develop appropriate climate change response and communication policies according to specific circumstances.

More attention should be paid to study of climate change communication in rural areas, especially remote poverty-stricken areas. Rural areas are most vulnerable to climate change and have accumulated a lot of valuable folk wisdom and local knowledge about climate change. Climate change communication for rural areas should be subject to a two-way design. On the one hand, we should inform farmers of the certainty of climate change through diversified channels and provide risk perception training service. On the other hand, more professionals should be invited to collect local knowledge of climate change accumulated in rural areas, to deepen the research, explore the logic and law behind it, and promote experience and communication with other developing countries, so as to make local knowledge play a greater role.

In order to mobilize more people to participate in effective communication, we should analyze the audience's discourse mode, psychological characteristics and thinking way.

4 Case Study

As a worldwide poverty alleviation and development agency, Oxfam focuses on helping vulnerable poor groups respond to climate change. However, repeated emphasis of this point will make the donors and the general public feel tired in time. In order to better mobilize the public, Oxfam designs different frameworks, targets at different audiences, and deeply interprets the relationship between climate change and poverty, taking into account sustainability and timeliness. In March 2014, IPCC released the fifth climate change assessment report, saying that "the impact of climate change on food security and global hunger will be more serious than expected, and will come earlier than expected". In response, Oxfam expounded the harms of climate change from the perspective of food security, planning and launching the "Food and Climate Justice" campaign for more than 40 countries, and emphasizing the negative impact of climate change on food security and small farmers, especially vulnerable groups such as women and children. Oxfam submits research reports to the government for the purpose of advocacy. For example, it published the research report entitled "Heat and Hunger: How to Prevent Climate Change from Affecting Anti-Hunger Action", which analyzed ten important factors affecting the country's supply of food for people and called for the attention of government. Oxfam conducts the whole-process supervision of private sectors such as multinational corporations relying on the power of consumers, requiring them to disclose emission data, and taking active and effective actions to reduce harmful emissions and protect poor farmers, to ensure that everyone has sufficient and high-quality food. For consumers, Oxfam has designed the slogan of advocating food safety and low-carbon life, and promoted it in supermarkets, shopping malls and other occasions. A variety of events hosted by Oxfam convey the same signal: Climate change has become one of the biggest challenges we face in fighting against hunger and poverty. Extreme weather events and seasonal changes are destroying harvests, pushing up food prices and lowering food quality, which will affect farmers' choices of farming and dwelling, and also the choices of urban residents. Precisely due to the targeted communication based on clear core information, six countries have accepted Oxfam's suggestions in only one year, beginning to focus on food security. Five multinational companies including Nestlé and Pepsi have agreed to publicly release data of emission, and nearly 200 million people around the world have participated in the "Food and Climate Justice" campaign.

In short, in order to achieve collaborative governance and strive for the largest win-set at the domestic level, all stakeholders including scientists should take action and play a positive role in the dynamic feedback system. Only in this way can China play a positive role in global climate governance, so as to further stimulate the enthusiasm of all stakeholders in China to address climate change, truly form a "dual-layer, multi-dimensional" collaborative governance structure, and promote the win-win interaction in China in the post-Paris era.

Chapter 7
New Stage of "Dual Transition" (2015–2018)

Global governance is a dynamic development process, and climate change, as one of the important agendas of global governance, is subject to the dual impact from both the international political and economic pattern. In the context of the global economic depression caused by the global financial crisis, hidden crises emerged in the wake of a series of events, including the new contradiction triggered by the European refugee crisis in 2016, the retrogression of the European Union's integration process following the U.K. divorce with the European Union and social bipolarization exposed during the U.S. president election. The "reverse globalization" thought and protectionist tendency picked up, great uncertainties existed in policy trends of major economies and their spillover effects, and unstable and uncertain factors obviously increased. On the international and domestic game fields, Europe and the U.S. obviously shifted their focus of attention to domestic affairs, which reduced the importance and urgency to participate in global governance, which has accordingly affected their willingness to play a leading role in global climate governance.

On November 7, 2016, the COP22 took place in Marrakech, Morocco, a North African country, and Trump, a skeptic of climate change, won the U.S. president election two days after the opening of the conference, which added new uncertainties to global climate governance.

Our core view in this chapter is that the global climate governance has entered into a new stage of "dual transition" after the COP22 in Marrakech, mainly evidenced by two facts: First, the leadership structure of the global climate governance need to be rebuilt because the former China-U.S. joint leadership on climate collapsed; Second, the emissions reduction mode has changed from the "top-down" mode to the "bottom-up" mode. The new stage is both an opportunity and also challenge for China, which should explicitly establish the strategy for climate change communication and governance.

1 "Dual Transition" Stage of Global Climate Governance

Since the 1970 s, the global climate governance has yielded a series of important periodical fruits, including the United Nations Framework Convention on Climate Change, the Kyoto Protocol and the Paris Agreement after a tortuous development for more than 40 years. In particular, the UNFCCC defines the basic framework of the global climate governance system, and the Kyoto Protocol and the Paris Agreement are two milestone documents with legal effect in the history of global climate governance. As to the top-level design of the governance structure, the European Union played a leading role in the era of the Kyoto Protocol, and China and the U.S. played an essential role in validating the Paris Agreement. As to the emissions reduction mode, the Kyoto Protocol defines the "top-down" mode whereas the Paris Agreement selects the "bottom-up" mode. Further dynamics of the international political situation will further strengthen the uncertainties of global climate governance and plunge the leadership structure and emissions reduction mode facing dual challenges, and "dual transition" will become a typical characteristic of this stage.

1.1 Transition of the Leadership

As a typical global public product, climate change involves various areas of social and economic development. In the book *The Logic of Collective Action*, the U.S. scholar Mancur Olson explains the core of the collective dilemma theory, that is, rational individuals seeking self-interests will not take an action to realize their common interests or interests of the group, unless one group has very few people or there are mandatory or other special means that promote individuals to act for their common interests. From the Kyoto Protocol to the Paris Agreement, the global climate governance has made important progress and successively overcome the collective dilemma, thanks to the concerted efforts of the international community. Leadership plays a decisive role in international cooperation and negotiation, particularly overcoming barriers to attain international agreements and establish international consensuses. Leadership is traditionally considered as an asymmetric influencing relation, namely, one party guides the actions of other actors to attain one objective. It can provide a mode other parties are willing to imitate, thereby reducing the uncertainties of collective actions.

Looking back to global climate governance, particularly the process of international climate negotiation, we can find the leadership has always changed from one to another.

In December 1997, 140 countries signed and adopted the Kyoto Protocol, which defines the obligations of emissions reduction in a top-down manner, that is, all developed countries should reduce the omissions of six greenhouse gases, including carbon dioxide, by 5.2% from 1990 to 2010, and the developing countries assume no obligations. The protocol must be approved by the countries contributing 55% of the greenhouse gas emissions in 1990 before taking effect. The Clinton Administration signed the Kyoto Protocol, but after assuming office, the Bush Administration declared to withdraw from the protocol in 2001, which delayed the validity of the document. At the end of 2004, Russia approved the Kyoto Protocol, which set the key condition for the validity of the Kyoto Protocol and assured its formal validation one year later. If leadership meant taking the lead to take substantive emissions reduction measures, the U.S. and China would be both reluctant to play the leading role in global climate governance in the era of the Kyoto Protocol. In particular, the withdrawal of the U.S., then largest emitter and economy, further reduced the efficiency of the mandatory "top-down" emissions reduction mode stipulated by the Kyoto Protocol. The EU played a big role in setting the agendas during the Kyoto Protocol time. After the U.S. withdrawal, EU propelled and led the work to validate the Kyoto Protocol by submitting proposal drafts and seeking two-way compromise and cooperation with the developed countries and the developing countries. The EU also took active measures to influence the domestic approval by other key countries and promoted them to actively perform the protocol after validation. However, the climate leadership of the European Union contracted, according to the key actor survey performed by the researchers in 2008–2011 (as shown in Table 1). In particular, during the COP15 in 2009, the monorail process proposed by EU was not realized but strongly opposed. "It has been directly marginalized" (Parker, Karlsson & Hjerpe, 2015).

After the EU lost the climate leadership, the global climate leadership first incurred a vacuum of leadership, which rendered a fragmentation trend. Around the COP21 in Paris, China and the U.S. met the expectations of all countries and joined hands to shoulder the leadership of global climate governance. They have finally promoted the formation of a global climate governance mode featuring coordination by China and the U.S., multilateral governance and extensive participation of diverse stakeholders.

From the EU to China and the U.S., the global climate governance leadership completed a round of transition. The attitudes of China and the U.S., the largest economies and emitters in the world, determine the intensity and trend of global

Table 1 Analysis of perceptions of key actors' climate leadership (2008–2011)

Analysis of climate leadership	COP14 (2008)	COP15 (2009)	COP16 (2010)	COP17 (2011)	Deviation (2008–2011)
European Union	62	46	45	50	−12
China	47	48	52	50	+3
Group of 77	27	22	19	33	+6
The U.S.	27	53	50	42	+15

Quoted from Charles F Parker, Christer Karlsson and Mattias Hjerpe, "Climate Change Leaders and Followers: Leadership Recognition and Selection in the UNFCCC Negotiations", International Relations VoL 29 No. 4, 2015, pp. 434–454
(*Note* Sample size = 1571)

climate governance. In November 2014, September 2015 and March 2016, Chinese and the U.S. leaders successively promulgated three joint statements on climate change. On December 13, 2015, 196 contracting states reached the Paris Agreement, which has made the institutional arrangement for global common response to climate change after 2020. The COP21 in Paris has achieved a generally accepted diplomatic success, given huge difference in national situation, interest and perception. On April 22, 2016, the high-level signing ceremony for the Paris Agreement was held at the United Nations Headquarters in New York, and the leaders from a total of 175 countries, including China and the U.S., signed the agreement, which has created the record for the largest number of the countries that have signed an international agreement on the first opening date. On September 4, 2016, China and the U.S. took the lead to approve the Paris Agreement before the G20 Summit and increased the ratio of the emissions of the countries joining the Paris Agreement to global total emissions to nearly 40%. This move has played an essential role in promoting the conclusion of the Paris Agreement with historic significance as well as its final validation and implementation. On October 5, the Paris Agreement met two validation conditions, that are, 55 contracting states join the agreement and contribute more than 55% of global total greenhouse gas emissions. On November 4, the Paris Agreement formally took effect (see Table 2).

The validation of the Paris Agreement, which is the most complex, sensitive and comprehensive result of climate negotiation so far, within such a short time, has reflected the strong determination of all countries to take global actions in face of the climate change and fully displayed the appeal of China and the U.S. after they joined hands.

On November 7, 2016, the COP22 opened in Marrakech. It is the first conference after the formal validation of the Paris Agreement, and the results at the conference will deliver a decisive influence on the climate negotiation in several years to follow and effectively tested the validity and credibility of the commitments made by different parties. The results of the U.S. president election came out, and Trump, a skeptic of climate change, won the U.S. general election two days after the opening of the conference.

During the presidential campaign, Trump denied climate change and believed global warming was a "conspiracy". Refusing to believe climate change related to human activities, Trump said that if elected, he would ask the U.S. to withdraw from the Paris Agreement and stop all financial supports for the climate change project of the United Nations. On the first day after he assumed office, the official website of the White House immediately deleted all contents related to climate change and published the U.S. First Energy Plan (White House, 2017). Trump also nominated an oil company official to serve as the State Secretary and nominated a skeptic of climate change to lead the United States Environmental Protection Agency. Thus, he was accused of "asking a firebug to put out the fire" (Lin, 2017). It is certain that the U.S. government will no longer continue the positive climate policy made by the Obama Administration, which will terminate the landscape of joint leadership featuring China-U.S. collaboration established by the Paris Agreement. The

1 "Dual Transition" Stage of Global Climate Governance

Table 2 Key events for formal validation of Paris agreement

	Major event
November 2014	Chinese and the U.S. leaders issued the first China-U.S. Joint Statement on Climate Change
September 2015	Chinese and the U.S. leaders issued the second China-U.S. Joint Statement on Climate Change
March 2016	Chinese and the U.S. leaders issued the third China-U.S. Joint Statement on Climate Change
December 13, 2015	196 contracting states reached the Paris Agreement, which has made the institutional arrangement for global common response to climate change after 2020
April 22, 2016	The high-level signing ceremony for the Paris Agreement was held at the United Nations Headquarters in New York, and the leaders from a total of 175 countries, including China and the U.S., signed the agreement, which has created the record for the largest number of the countries that have signed an international agreement on the first opening date.
September 4, 2016	China and the U.S. took the lead to approve the Paris Agreement before the G20 Summit and increased the ratio of the emissions of the countries joining the Paris Agreement to global total emissions to nearly 40%. This move has played an essential role in promoting the conclusion of the Paris Agreement with historic significance as well as its final validation and implementation.
October 5, 2016	The Paris Agreement met two validation conditions, namely 55 contracting states join the agreement and contribute more than 55% of global total greenhouse gas emissions
November 4, 2016	Paris Agreement formally took effect

Source Prepared by the author

U.S. sudden turnaround has caused global climate governance to suffer a passive leadership transition.

Looking back to the handover process of climate governance leadership as above, we can find that China was only a negative follower during the first handover of climate leadership. During the second handover of climate leadership, China changed from a follower to a leader. China's attitude has changed for two reasons.

First, the Copenhagen Negotiation in 2009 has delivered a strong impact on China. The Copenhagen Negotiation overtook the previous negotiations in scale and level, and the agendas became more complicated. The Chinese Government paid great attention to the negotiation, and Chinese leaders took the initiative to mediate the negotiation. However, 190 contracting states participating in the negotiation generally opted to hold fast to the upper limit and squeeze the space for compromise and negotiation and refused to make any compromise in relation to their interests. Finally, the negotiation failed to achieve an expected result but only ended with an agreement without legal binding force. *The Guardian* published at the first time an article blaming China and other countries for "wrecking" the Copenhagen Negotiation (LYNAS, 2009). The U.S. took advantage of the situation and made China the scapegoat that should be responsible for the failure to reach a legal document during Copenhagen Negotiation. China was then imposed a negative label of the "kidnapper of Copenhagen Negotiation" and suffered a serious negative impact on its international image. This is the external stimulus that has promoted China to change the attitude towards participating in global climate governance.

Second, from the perspective of the two-level game, the attitude change of one country at the international level and the change in the domestic policy environment influence one another. Judged from the internal cause, China's domestic economic development depended on heavy industries with coal and steelmaking as two pillars in 2000–2013, and compared to economic development, climate change and environmental protection were not put on the priority list of the policymakers. Therefore, the Copenhagen Negotiation displayed rigid attitudes and stances in various aspects without much space of compromise. Seeing pure pursuit of rapid economic development has caused environmental damage and depletion of natural resources, Chinese Government has upgraded the development of the low carbon economy to the national strategy level in the Twelfth Five-year Plan promulgated in 2011, and stressed the ecological civilization featuring green and low carbon is the common development trend of mankind. China has since then shifted to the low carbon development route, which echoed the trend of global climate governance and became the internal momentum for China to make an active contribution to the Paris Agreement and exert the climate leadership hand in hand with the U.S.

1.2 Transition of Emissions Reduction Mode

The Paris Agreement signifies the transition of the global emissions reduction from the "top-down" mode to the "bottom-up" mode, which will formally come in 2020. Therefore, the period from 2015 to 2020 will be the transitional period of the global climate governance mode.

The Kyoto Protocol valid in 2005 allocates mandatory emissions reduction indicators with the legal binding force to the developed countries, which kicked off the "top-down" climate governance. The mode features strong legal binding force, implementation of rigorous compliance mechanism and uniform accounting rule and defines stringent measurement, reporting and validation rules to assure transparency. However, it also has the disadvantages of the difficulty in reaching consensuses among various stakeholders, slow progress and low efficiency.

Compared to the Kyoto Protocol, the biggest highlight of the Paris Agreement is to adopt the "Intended Nationally Determined Contributions" (INDC) regime, which allows various countries to make emissions reduction and other commitments based on respective economic and political situations. This "bottom-up" mode has replaced the "top-down" mandatory mode established during the Kyoto Protocol period and permitted various countries to determine respective climate change response actions based on respective situation, capacity and development stage and make "differential" contributions while "commonly" submitting the INDCs. It is a dynamic development of the principle of "common but differentiated responsibility" defined by the UNFCCC and stresses great inclusiveness. It has inspired comprehensive participation of all stakeholders, including civil society, to the biggest extent.

Before the opening of the Paris Conference, 188 countries submitted respective INDC plans to the Secretariat of the UNFCCC. To assure the realization of the objectives, the Paris Agreement has introduced the update mechanism with global review as the core and every five years as one period, which has supplemented the disadvantages of the global climate governance system in regular update. Therefore, the Paris Agreement is an important milestone and has opened up the new "bottom-up" mode based on the INDC regime supplemented by five-year review.

The Paris Agreement has made a new framework arrangement for the global climate governance mode, which is a "pragmatic compromise" and can satisfy basic expectations of different parties to the biggest extent. The agreement is ambitious to control the temperature rise by no more than 1.5°. Yet, the overall design lacks global action objectives, explicit time to peak, emissions reduction schedule and concrete plan to cancel the fossil energy subsidy. Moreover, it also doesn't provide concrete requirements and regime details to attain the objectives. No party has the experience in respect of how to measure, supervise and implement the INDC regime and all parties have to test the water together. Due to the strong external impact from Trump's election, these inherent risks concealing in modal transformation have not attracted adequate attention.

First, the INDC objectives of different countries are defective in terms of content design. Considering political and economic development situations at home and abroad, it remains rather uncertain whether the countries setting the objectives through the domestic procedure can accomplish quantitative emissions reductions on schedule. Besides, some developing countries with limited capacity have hired foreign consultancies with funds from international organizations to help them set the objectives, so they lacked the sense of owning these objectives at the very beginning. Other developing countries have mentioned they will demand corresponding financial and technological supports to make their contributions in their objectives. Fund and technology remain the difficulties of the international climate negotiation. Thus, we can imagine these countries will encounter big challenges on the way to attain these objectives. Besides, as there are no uniform standards, the INDC objectives submitted by different countries vary a lot in respect of methodological selection, baseline setting, objective design and other aspects of the calculation of emissions reduction. Thus, it is an issue how to attain the INDC objectives based on the tight schedule. The report of the United Nations Environment Program(UNEP) points out that the emissions reductions contributed by various parties are still insufficient to attain the objective of controlling the temperature rise below 2°. Even if all these objectives come true, the temperature will rise by about 3° by 2100.

Second, as to the constraint mechanism, the Paris Agreement emphasizes "moral restraint", which depends on international supervision and assessment institutions. Such restraint mechanism is an internal restraint, and behaviors of governance entities are active, positive and voluntary to a big extent, and the mechanism also doesn't require the states to make corresponding domestic laws to assure the realization

Table 3 Comparison of execution mechanisms for paris agreement and kyoto protocol

Name	Kyoto protocol	Paris Agreement
Mechanistic innovation	Clean development mechanism, joint performance mechanism and emission trade mechanism	INDC mechanism
emissions reduction requirement	Mandatory emissions reduction	INDC
emissions reduction mode	Top-down	Bottom-up
Constraint mechanism	Mandatory legal effect	Moral restraint
Constraint target	Developed countries	All contracting parties
Obligation assignment	50–50 to South and North	Responsibility sharing
Penalty mechanism	Yes	No

Source Prepared by the author

of respective objectives, meaning there is no domestic legal basis to assure the realization of the intended nationally determined objectives. The Paris Agreement also defines the "global review" concept, which, however, needs further discussion in terms of operating method, particularly effectiveness. The "top-down" mandatory emissions reduction mode established by the Kyoto Protocol is unsuccessful, because it hopes to set mandatory requirements with legal binding force, which has been opposed by the developed countries. The current "bottom-up" mode has drawn the earlier lessons and stressed initiative and voluntariness to assure the maximum participation. However, as there is no corresponding legal restraint, the mode may be refused and delayed anytime, thereby sliding to another loose extreme situation and causing the failure of the emissions reduction mode. Therefore, mechanism exploration and implementation of details will be very important during the transitional period. How to strike a balance between the two modes and effectively practice global governance during the transitional period is a key issue for the international community (see Table 3).

2 Opportunities and Challenges for China at the New Stage of "Dual Transition"

We can know from the above analysis that after the COP22 in Marrakech, global climate governance has entered the new stage of "dual transition", including handover of leadership and transformation of emissions reduction mode. The new stage means an increase of uncertainties, which will come as both opportunities and challenges for China.

Judged from the level of leadership transition, China will embrace opportunities, because the international community generally expects China to assume responsibilities and the experiences of global climate governance have the strategic values of contributing to the Belt and Road Initiative. Corresponding challenges will come from how to manage the expectations of the international community and actively explore a leadership route with Chinese characteristics in the complex international environment. Besides, the spillover effect of climate governance has not been generally accepted and also lacks corresponding theoretical framework.

Judged from the transition of the emissions reduction mode, China will embrace opportunities, because the NDC objectives have been ideally satisfied and also improved in terms of methodology and transparency. China will also encounter challenges, because if China wants to further contribute to the overall process of global climate governance on the basis of domestic achievements, it should help other countries attain the objectives while assuring the realization of its own objectives. This will involve assessing inherent details of the mode, implementing assessment, supervision and other institutional arrangements, and leading the compliance with these arrangements by example. Radical implementation will possibly exhaust China's capacity and affect both strategic visions of domestic economic restructuring and profound participation in global governance.

2.1 Opportunities

Judged from the perspective of leadership transition, the change in the U.S. presidents reflects political uncertainties hiding in the global climate governance system built by the Paris Agreement, and its effectiveness depends on political intents of national leaders to a big extent. After Trump assumed office, the leadership structure forged by China and the U.S. hand in hand is at the verge of collapse. When the U.S. is possibly "absent", China would like to continue to exert the leadership, which satisfies the general expectation of the international community.

First, the U.S. "withdrawal" has brought great uncertainties to global governance, and the international community cherishes "general expectation" that China will exert the leadership in a number of global governance issues, including climate change response (Lmighran, 2016). On January 17, 2017, Chinese President Xi Jinping attended the World Economic Forum in Davos, Switzerland, and theme of the summit was "Responsive and Responsible Leadership". It is the first time that Chinese national leader attended the World Economic Forum. At the opening ceremony, President Xi Jinping delivered the keynote speech entitled "Jointly Shoulder Responsibility of Our Times, Promote Global Growth", and noted, "We should adhere to multilateralism to uphold the authority and efficacy of multilateral institutions. We should honor promises and abide by rules. One should not select or bend rules as he sees fit. The Paris Agreement is a hard-won achievement which is in keeping with the underlying trend of global development." (Xi, 2017). During the speech, he also emphasized that "As long as we keep to the goal of building a community of shared future for mankind and work hand in hand to fulfill our responsibilities and overcome difficulties, we will be able to create a better world and deliver better lives for our peoples". This is a high-profile response of Chinese leader to the global governance and international order problems concerned by the international community, including climate change, which explicitly expresses China's positive intent to adhere to multilateralism and abide by the Paris Agreement. Compared to China's earlier attitude of "prudentially active and sometimes nervous" towards global governance issues, China has this time explicitly voiced the intent to adhere to multilateralism and global governance. When the global governance faced significant uncertainties, the speech gave a shot on the arm of global governance, won positive comments from the international community and also won further expectation of different parties about China's role and responsibility. The Washington Post's commentator thought that if we look back to the past 5 or 10 years, we can say it is truly a turning point, and China has taken one more step towards the role of global leader. *The Guardian's* article thinks that President Xi's speech has strongly defended economic globalization, safeguarded the Paris Agreement and displayed China has the intent to play a more important role on the international stage. Thus, if China continues to exert the leadership, this will both meet the expectation of the international community and also represent an important external condition to help China to develop towards a low-carbon and sustainable direction.

Second, if China actively seizes the opportunities during the transitional period of leadership, this will help perform more macroscopic strategic layout on the basis of global climate governance experience, provide a mighty support for further implementation of the Belt and Road Initiative and contribute the global climate governance experience to the process of China's constructive role of global governance. In 2013, Chinese President Xi Jinping announced the Belt and Road Initiative, which has navigated China to enter a new stage of opening up. Over past years, the initiative has bloomed, yielded fruits and earned global influences. Its implementation will not only influence global political and economic situations but also impact the energy, resource and environment of the countries along the Belt and Road. When speaking at the Legislative Chamber of the Uzbek Supreme Assembly on June 22, 2016, President Xi Jinping stressed we should build a Silk Road of "green, health, governance and peace", where "green" ranks the first and stresses the work to deepen environmental protection cooperation, implement green development concepts and strengthen ecological environment protection. It is worth special mention that the Belt and Road Initiative and the South-South Cooperation Framework at the international level overlap to a big extent, and most of the involved countries are impoverished developing countries and also the countries that suffer the most significant negative impacts from climate change. By implementing the Belt and Road Initiative based on the existing advantages in the climate change area, China can help the impoverished developing countries better respond to climate change, strengthen cooperation with the countries along the Belt and Road on this basis to satisfy actual needs of the developing countries to address climate change, and make due contribution to climate change response in the countries along the Belt and Road.

When looking back to the process of China's participation in global climate governance, we can find substantial changes have happened to China's role and influence from 2009 to 2015, and China changed from negative participation to active leadership and won general recognition from the international community. It is the first time and also the fastest that China has changed from negative position to the winning of say in the participation in global governance. Since the Belt and Road Initiative is an innovative exploration of global governance, then, if we take the initiative to sum up China's experience of participating in global climate governance and contribute to the Belt and Road Initiative, we can help China play a better role when it participates in more comprehensive global governance issues.

From the perspective of emissions reduction mode, during the realization of the INDCs under the bottom-up mode, China has taken the initiative at the first time and made the preparation to play a more constructive role by submitting the NDC proposal and publishing the assessment report. On June 30, 2015, China submitted the Enhanced Actions on Climate Change: China's Intended Nationally Determined Contributions, namely the First Two-year Update Report of the People's Republic of China on Climate Change. According to the report, China will reach the carbon emission peak around 2030, strive to reach the peak as early as possible, reduce the carbon dioxide emission of unit gross domestic product (GDP) by 60%–65% in 2030 compared to 2005, increase the proportion of non-fossil energy up to 20% in primary energy consumption, and increase the forest reserve by 4.5 billion cubic meters compared to 2005. Meanwhile, the report also stipulates the objectives of climate change response actions after 2020 as well as the roadmap and policy measures to attain the objectives. It is a specified action China has taken as a party to the Convention, and has also exhibited the resolve and attitude of Chinese Government that China will go the way of green, low carbon and circular development featuring growth transformation, energy transformation and consumption transformation to the domestic public and the international community.

In December 2016, Chinese Government submitted the first two-year update report pursuant to the provisions of the Paris Agreement. Data in the report show that China has attained the forest reserve objective by more than twice, reduced the carbon intensity by 97% and fulfilled another two objectives by 60%. In the report, Chinese Government has calculated and updated emission data with more scientific methods and added the calculation of 18 emission sources. Moreover, China has promised to further make the methods more scientific. The international community has seen that "China is proving its leadership in climate governance with tangible actions" (Rossa, 2017) from China's serial measures, including disclosure of more scientific data, disclosure of objective fulfillment and update of objective setting.

The two-level game theory stresses the leader of every country simultaneously plays two games. One game is the international negotiation table, and the counterparty is the negotiation representatives from other countries. The other game is the domestic negotiation table: The leader must balance the interests of different interest groups and obtain the largest winner set. At the same time, theory also stresses the analysis from the perspective of domestic-international interaction, rather than simple stack of international and domestic factors. Judged from the perspective of two-level game, the new stage of "dual transition" will be both opportunities and challenges for China. From the perspective of opportunities, China domestically has the intent and condition to continue leading climate change response, and externally, China has the support from the international community. If China seizes this opportunity to consolidate the climate leadership, it will both contribute to hard-won results of global climate governance and also contribute to China's diplomatic strategy as a constructive power nation.

2.2 Challenges

The international order and global governance issues are complex and diverse, and China also faces dual challenges at the global climate governance mechanism and mode level while embracing opportunities.

From the level of cooperation situation, leadership research stresses the supply-demand balance: leadership depends on who wants to become the leader, what he says and does, and his objective, strategy and roadmap. While China must also examine what followers expect and who can satisfy such expectation. The traditional leadership research only emphasizes "supply" but ignores "demand". Only supply and demand balance, can leadership become truly sustainable.

From the perspective of demanders, the international community generally acknowledges the key role China has played in and after the COP21 in Paris. This has also provided a good external environment for China to continue reinforcing climate leadership. Yet, it is a challenge for Chinese Government to precisely position the leadership with Chinese characteristics under the complex situation so as to both meet the expectation of the international community and highlight China's diplomatic style as a constructive power nation.

Besides, the level design of global climate governance is now being rebuilt, and China will need more collective wisdom to actively sum up the experience and contribute the experience to the Belt and Road Initiative and global governance in a more comprehensive sense while further carrying forward climate governance. Currently, the logical correlation between climate change and the Belt and Road Initiative should be combed, and in this regard, there are still very few researches and demonstrations based on the green perspective. "Green" is identified as a priority in the Belt and Road Initiative. Yet, it still lacks consistent action outline and plan in the dimension of top-level design and involves defining comparable baselines and guiding principles and researching diverse, fair solutions, assessment systems and action guides. The absence of theoretical research based on the "green" perspective doesn't match the implementation progress of the Belt and Road Initiative. Besides, it is scarcely demonstrated how to focus on the Belt and Road Initiative from the perspective of climate under the green development framework, extend climate governance to more comprehensive global governance and extend it to sustainable development goals. The absence of the top-level design theory will lead to errors of strategic choice. Therefore, refining theories in this regard will be the top priority, if China wants to more actively exert the leadership in global climate and related governance areas.

From the perspective of emissions reduction mode, a number of problems will happen during the transition from the top-down mode to the bottom-up mode, such as inadequate transparency and unsound assessment/constraint mechanism, while the international community is full of expectation of China. To satisfy the expectation means more responsibility, and to shoulder the responsibility, China should both have a strong political intent and take tangible actions to truly establish public trust. The formal validation of the Paris Agreement marks global climate governance will enter the performance stage after 2020. Performance capacity is the most important among all factors that influence performance (Zhang, 2016). China has gone the way to evolve towards low carbon, but it still suffers heavy burdens from economic development. To accelerate the green and low carbon transformation of economic development during the transitional period of economic growth slowdown and reinforcement of structural adjustment, China must realize the objective to control the intensity and aggregate of greenhouse gas emissions. Meanwhile, China should also carry out various measures and policies to reduce greenhouse gas emissions for the purpose of promoting pollution control and realizing a sustainable development. This is a "very important and systematic process of technical innovation and system change" (An & Zhang, 2016). Besides balancing its own capacity, China should also see the developed countries vary to some extent in terms of performance capacity, while the developing countries generally face inadequate capacity in fund, technology, regulation, policy and other dimensions. If China takes a blind-minded action without considering the performance capacity, it will possibly suffer a capacity overdraft anytime. Thus, China will face a host of internal and external challenges on the way to lead the global climate governance.

3 Strategic Choice

Based on the aforesaid analysis, China must make a long-acting strategic choice, explore the climate leadership system with Chinese characteristics, perform all-round overall planning and work out Chinese solutions for global climate governance to change challenges into opportunities, convert stress into momentum, advance the global climate governance, and further contribute to the sustainable development.

3.1 Strategic Choice of China's Climate Governance

The climate leadership with Chinese characteristics can be interpreted in three dimensions.

First, China's development stage determines the climate leadership with Chinese characteristics should not be omnipotent but should seize specific advantaged areas to make major breakthroughs.

Second, unique leadership no longer suits the current climate governance, judging from the complexity of climate governance and the multi-polarization trend of world development. Partnership should also become an important mark of China's diplomacy Therefore, the leadership with Chinese characteristics should not be exclusive but be inclusive and reflect the concept of sharing and joint construction.

At the new stage of "dual transition", the most important is to hold fast to the strategy, seize external opportunities and consolidate existing advantages in climate governance so as to lay a solid foundation for further guiding global climate governance. In this sense, the strategy should be made after short-term, medium-term and long-term strategic considerations.

In the short term, to address dynamics of climate governance, China should continue to take an active part in the global climate governance system with existing advantages. China's advantages lie in the climate negotiation under the United Nations framework and the south-south climate cooperation. Next, the climate negotiation will enter the technical discussion stage, which will emphasize the realization of the INDC objectives submitted by different countries. The Chinese Government is confident to accomplish the INDC objectives and will take more active measures in concrete performance of the objectives and present its own solution. In contrast, the south-south cooperation in climate change will encounter a tough task—it must overcome mechanism and system problems during the operation of the South-South Climate Cooperation Fund, refine the management system and build a long-acting assessment mechanism. Also, the mechanism must actively enlarge funding sources and inspire the enthusiasms of diverse parties, including private sectors, to participate in the process. This will be very important to exert China's funding and technical advantages in climate change response and win the moral and public opinion support from vast developing countries to jointly realize a sustainable development.

In consideration of the uncertainties at the current stage, the medium and long-term strategic considerations will be more important to maintain the strategic poise. In the medium term, China should carry forward new climate leadership building, and perform different scenario and potential analysis of possible partners based on the analysis of types of climate leaderships once exerted by the European Union and the U.S. The Chinese leaders have emphasized the principles of "inclusiveness, cooperation, mutual trust and win-win" on several occasions, and instructed by these principles advancing with the time, China should actively interact with the international community and explore a new leadership pattern, inclusive, open and mutually beneficial, with Chinese characteristics. In particular, the China-U.S. relationship is now the most important bilateral relationship with profound foundation and extensive interests in the world now. Despite some disputes and contradictions, both parties still actively explore constructive relations in the overall direction to seek similarities out of dissimilarities, cooperation and win-win. Considering these points,

when designing the new climate leadership system, China should leave a strategic regression space for the U.S., which will demand more courage and wisdom.

3 Strategic Choice

In the long term, China should complete the top-level design of global governance beyond the climate change issue. Under the new global governance system, China will more play a leadership and coordinating role, rather than a bossy leader in the traditional sense. Only by doing this, can China inspire the enthusiasms of different parties to participate in the system and solve the collective dilemma with collective force. After systematically combing the experience and lessons related to climate change, China should act at the right time to connect climate change response and realization of sustainable development goals, duplicate the experience of climate governance to the broader global governance issues, including the Belt and Road Initiative, and boost the realization of global sustainable development goals (as illustrated in Fig. 1).

In a word, global climate governance has entered the new stage of "dual transition", including governance mechanism leadership and mode, after the COP22 in Marrakech. The governance mechanism has presented a transition of leadership, and the emissions reduction mode has evolved towards the "bottom-up" mode. China should both embrace opportunities and also encounter challenges behind them during the transitions at the two levels. China can make more constructive contributions to global governance by performing strategic design and converting challenges into opportunities.

Over the past years, global climate governance is the forefront where China has participated in global governance and gained rich experience and lessons in winning the speech right and public trust. The grand context behind the "dual transition" of climate governance is the profound adjustment of the international system and international order. Live up with the opportunities and challenges from the new stage of "dual transition" and make a correct strategic choice—this will be another exercise for China to participate in global governance. By maintaining the strategic poise after identifying the characteristics of the current stage, China can more actively seize opportunities, meet challenges, accumulate valuable experience in participate in global governance in depth and actively contribute to the realization of sustainable development goals of mankind.

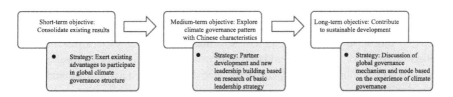

Fig. 1 Strategic design for China's participation in global climate governance at new stage. *Source* Prepared by the author

3.2 Strategic Response for China's Climate Change Communication

In the context of global climate governance change, China should also design the corresponding strategy for climate change communication based on the short, medium and long-term strategic considerations of climate governance.

First, China should coordinate the domestic and international public opinion circles, solidify the international and domestic resolve to respond to climate change as well as the concept of international climate cooperation, assemble consensuses and consolidate existing climate leadership.

In face of the international audience, China should actively convey the message that China is firm to address climate change through various channels and in various forms so that the international community will see China's efforts, including progress and periodical achievements of INDC objectives as well as Chinese public's participation in climate change response. In face of the domestic audience, China should timely communicate positive actions of the international community to address climate change so that the domestic public will feel the actions and determination from other countries, thereby supporting Chinese Government to assume more climate leadership. For example, the YPCCC has persisted in the survey on the U.S. people perception of climate change for more than ten years. Just one month after Trump won the election, the project completed a poll, showing that "70% of the U.S. people support the restriction over building of coal-burned and thermal power plants, and 69% of the U.S. people support the U.S. government's performance of the Paris Agreement" (Leiserowitz, Maibach, etc., 2016). If these messages are timely reported by Chinese media, they will play a positive role to stabilize public opinions.

Climate financing is always a difficulty during international climate negotiation, because the developed countries are reluctant to provide substantial financial supports for the developing countries. The Paris Agreement confirms that after 2020, the developed countries should annually provide US$100 billion for the developing countries, and the new amount will be confirmed and continuously increased before 2025. According to the report of the Organization for Economic Cooperation and Development (OECD), the public finance support of the developed countries for the developing countries to address climate change was US$37.9 billion in 2013 and US$43.5 billion in 2014. The OECD also forecasted this figure will reach US$67 billion in 2020. The international community has greatly debated the fund amount in the OECD's report, and many countries have doubted the statistical method. Even if we don't consider the dispute in the composition of these public funds, these data are still far from enough to meet the lower limit of US$100 billion specified by the Paris Agreement.

The global climate financing has a huge gap. Because of monotonous information source, "inadequate climate financing" has been frequently reported by Chinese media, which has strengthened the public concern about the prospect of global climate governance. Yet, the domestic media have ignored that the international community has not stopped the pace on account of the slow pace of the developed countries in relation to climate financing and huge funds have flown to the climate area. In December 2016, the international climate financing was active, according to statistical data of the Climate Bonds Initiative. On December 12, Monash University of Australia offered the world's first climate bond of its kind. On December 14, the Brazil New Economy Forum opened and vigorously promoted the Sustainable Energy Fund (SEF) with a fund pool of US$144 million that aimed at guiding private sectors to make green investment and promote the growth of its domestic green bonds market. On December 15, Poland, a major coal producer, offered a national climate green bond in an amount of EUR700 million.[1] By actively collecting and reporting such information, Chinese media can both bolster domestic confidence and also provide new thoughts for climate financing.

Second, China should forge an agenda alliance with international media and NGOs, strengthen the supervision of different countries' fulfillment of INDC objectives and fully exert the social role.

The society-driven bottom-up emissions reduction mode lacks mandatory standards, so its implementation should be supervised through moral, public opinion and normative forces, which is also a new challenge for the objectives of the Paris Agreement. It is the inherent mission of NGO to promote the development and implementation of issues related to sustainable development. When the public are poorly aware of climate change, it is NGO that enlighten the public with various initiative actions so that they will realize the impacts of climate change, take positive actions to address climate change. In global climate governance, NGOs shoulder the mission of supervision, and play the role of monitor when participating in international climate negotiation to assure a force balance between the developed countries and the developing countries. They have also gained some professional expert in this process. The bottom-up governance pattern has provided a new space for NGOs, which will become more inspired to ally with media and government, set standards, refine mechanisms and perform the supervision responsibility. At the COP22 in Marrakech at the end of 2015, the German Government and the Moroccan Government jointly launched the "NDC Partnership" initiative, which will be performed by the World Resources Institute, an independent NGO, to provide information, knowledge, technical and financial supports for different countries to accomplish the INDC objectives.[2] Such cooperation has won the acknowledgment from the international community, enlightened the appearance of more partnerships and agenda alliances and played the role of public opinion supervision.

China's NDC objectives consider both mitigation and adaptation and explicitly set the roadmap from the present to 2020, 2030 and future. The objectives have provided a detailed implementation plan and won the reputation as a best practice globally. On this basis of strategic confidence, China may welcome international

---

[1] Data come from the official website of the Climate Bonds Initiative, http://www.climatebonds.net/2016/12/poland-wins-race-issue-first-green-sovereign-bond-new-era-polish-dimate-policy.

[2] For details, please see the website of the NDC Partnership www.ndcpartnership.org.

media and NGOs to supervise itself and set an example in the implementation of the NDC objectives to both boost the transformation of the climate governance mode and also credit the climate leadership.

Finally, China should actively sum up and deepen agendas and enlarge the width and depth of climate change communication.

Based on the perception of the synergy between climate change response and the United Nations Sustainable Development Goals, China should actively sum up the experience of climate governance and climate change communication, identify profound relationship between climate change and sustainable development. on this basis, China can select a framework in the language that is easily understandable to the international audience to make a collaborative communication effect, thereby enlarging the width and depth of climate change communication.

The condition precedent for deepening agendas is to fully understand current agendas. China should adhere to principles for climate change communication in the context of global climate governance dynamics: First, adhere to the principle of "common but differentiated responsibility". The principle of "common but differentiated responsibility" is the basic principle stipulated by the UNFCCC, an important element of the international climate governance mechanism and also the baseline the developing countries represented by China have always adhered to. When applying this principle, the Paris Agreement adds the dynamic factor, and some researches think the Paris Agreement strengthens the common responsibility but weakens the differentiation. Media should pay attention to this point when communicating related agendas. Second, the US President Trump denies climate change. However, consider the U.S. importance in the international order and global climate governance, media should conduct climate change communication based on a long-term perspective, reserve a strategic leeway for the U.S. during the language choice and also win an active, flexible strategic space for China.

To enlarge the width and depth of climate change communication, we should not purely rely on media but count on or actively establish the stakeholder partnership, and particularly pay attention to the force of enterprises. Enterprises are the principal force of energy saving and emissions reduction. Moreover, they have direct economic transactions with the public. Thus, if enterprise managers truly pay attention to climate change at the strategic level, enterprises will play a special role in climate change communication. During the COP22 in Marrakech, Ms. Qiaonv He, a Chinese female entrepreneur, declared to donate 100 million yuan to build a special climate change fund. The donation is the first private fund around the world after the formal validation of the Paris Agreement. It has given a shot on the international community persisting in climate governance as it was declared during the COP22.

To sum up, it is of special significance to strengthen China's climate change communication in the context of global climate governance, and China should select the proper response strategy after reviewing the experience and lessons to deal with various changes and win the strategic initiative.

Chapter 8
Conclusions and Outlooks

On the basis of the theoretical review and case analysis in the aforesaid seven chapters, we will assess the values of the "dual-layer, multi-dimensional" research framework and applying the related theoretical tools, sum up China's experience of participation in global climate governance in the past eight years and look ahead to China's future of climate change communication and governance.

1 Major Conclusions

1.1 Theoretical Levels: The "Dual-Layer, Multi-dimensional" Research Framework

Based on the theoretical review and case analysis in the previous chapters, the "two-level, multi-dimensional" panoramic research space is built up. Two levels mean the international and domestic dual-level game situation. Multiple dimensions mean simultaneously examining the participant, time and other different dimensions. With this framework, dynamic changes of one country participating in such domestic and international game can be tracked.

Global governance is the process that actors on the international stage take collective actions to solve common issues facing by the globe. During the research on global governance levels, scholars have gradually made clear the governance route and generally accepted that multidimensional and multi-element cooperative governance is the most significant global governance pattern. The transition of the global governance pattern from the state-centrism governance to the multi-element and multilevel collaborative governance is both a reflection of the global governance reality and also an expression of the deepening of global governance. It will not only help drive global governance to become more democratic, equitable and inclusive but also power the governance system to achieve dynamic and benign development. Moreover, the new pattern will both enhance the effectiveness of global governance and enhance the synergy effect of overall governance.

From the perspective of multiple dimensions, Robert O. Keohane and Joseph S. Nye, Jr. divide global governance into three vertical levels: supranational (including multinational corporations, intergovernmental organizations and nongovernmental organizations); national (including companies, central government of the state and democratic society) and sub-national (local companies, local governments and local democratic societies). Later, some scholars have enriched the vertical governance levels with the horizontal levels by dividing governance into the vertical governance that is committed to the high-level governance and the low-level governance and the horizontal governance that is dedicated to the non-state governance by multiple entities. From the perspective of multiple elements, the deepening of globalization has led to the decentralization of authority and the transfer of the central power in two directions: First, transfer to other levels in the vertical direction; second, transfer to non-state actors in the horizontal direction.

The old global governance pattern led by the U.S. and characterized in hegemony no longer adapts to the progress of the times and should be replaced by the new pattern featuring mutual dependence, equality and mutual benefit, and the future direction should be the pattern with the core of joint governance by both developed countries and developing countries as well as state actors and non-state actors.

The multilevel and multi-element governance pattern is an ideal mode, and different countries should have different ways to attain the pattern based on different development stages. From the global financial crisis in 2008 to date, China, India and other emerging economies have risen in global governance as a reality, and in particular, China has caught the eyeballs of the entire world with its active performance in global governance. The emerging economies generally face the question of exploring their own ways to participate in global governance based on different situations and development stages, rather than simply duplicating the practices of Europe, U.S. and other western democratic countries.

China will still have a long way to go before reaching the "multidimensional and multi-element" ideal governance pattern. However, very few scholars have even observed this gap between ideal and reality. This is because the academic circle generally faces a dilemma—they can designate the ideal other side built on theories, but they can't clearly see the way that leads to the ideal due to the lack of long-term profound involvement.

Noticing this problem, I degrade the research dimension to the national level and focus on the transformation of China's role in global climate governance. Starting from the national level but ranging across more than one level, I have tracked the work in the field Chinese stakeholders engaged in for consecutive years and built a "dual-layer, multi-dimensional" research space (see Fig. 1), thereby building a bridge between the reality and the ideal.

The "two-level" can assure China shares basic concepts and consensuses with the international community and also consider China's actual development stage. China is now walking on the way to go global and take an active part in global governance.

1 Major Conclusions

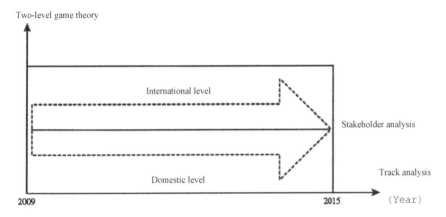

Fig. 1 "Dual-layer, multi-dimensional" research framework (Take 2009–2015 for example). *Source* Prepared by the author

The international level is divided into global, regional and other segmented levels, but when facing the issues at any level, China will still mainly focus on how to balance these issues with domestic development. Thus, the two-level game remains an effective theoretical tool to address concrete governance issues. Compared to the "multilevel" framework, the "two-level" framework can correspond to strategic objectives more clearly.

Stakeholders under the "multi-dimensional" framework are individuals or groups that will influence the realization of climate change governance objectives and include state actors and non-state actors. State actors and non-state actors are classified from the perspective of actor's identity and contain the tendency of organizational standard. Stressing contributions of non-state actors is one of the basic attributes of global governance that has risen since the 1980s. Based on the stakeholder theory, the "dual-layer, multi-dimensional" framework stresses all entities are equal in face of common interests and joint governance by stakeholders and represents the practical mode advocated for global governance. The stakeholder theory can help researchers jump out of the power framework to identify the governance participants in a more comprehensive manner.

As a process of common participation of diverse actors in multiple levels, global governance demands top-level design and needs to strengthen macroscopic planning and coordination. At the same time, it also involves building a more effective fundamental system, which is an important way to march towards global governance. Global climate governance is always dedicated to adopting a global constitution with general legal binding force and has eventually reached the Paris Agreement. However, The U.S. withdrawal has dealt a heavy blow on the implementation of the Paris Agreement, which has urged people to review the effectiveness of the top-down governance pattern again. Meanwhile, the "minor multilateral" climate system between different states, public-private partnership and climate cooperation system are also making a positive development. The "dual-layer, multi-dimensional" research framework starts with the national level and explores changes in international and national strategies of government, media, nongovernment organizations and other stakeholders. The framework can integrate the latest practical progress into the research framework in a timely manner and shore up the fundamental system infrastructure for global climate governance.

The transformation of the global system has made it more imperative to reform global governance, while effective global governance also demands systematic innovation. The "dual-layer, multi-dimensional" research framework provides an insight into global climate governance at the national level and also points out a realistic roadmap that will lead to the ideal global governance pattern.

1.2 Practice Level: China's Route Choice

Since the outbreak of the global financial crisis in 2008, the global governance system and decision-making mode can no longer adapt to the complex and diverse new situations. The international community has called to reinforce global governance and kept strengthening the consensus of reforming and innovating the global governance mechanism. In this situation, the global governance system has entered a reform period, and are experiencing creation, modification and even rebuilding to varying degrees.

In 2009–2015, historic progress was made in global climate governance, and China played an essential role in this process and won global acknowledgement with its contributions. We can see from the analysis in Chap. 5 that when dealing with global climate governance, Chinese stakeholders have made significant changes of the concepts, systems and orders for implementing concrete climate change communication and governance at both international and domestic levels.

We can clearly find that under the "dual-layer, multi-dimensional" research framework, China has benefited from the two-level strategic concept of "coordinating international and domestic overall situations" and promoting the multi-element cooperation in practice. After learning the lessons in Copenhagen, the three stakeholders—Chinese Government, media and NGOs—have actively adjusted their two-level strategies to adapt to each other under the common objective. They finally set up the strategic partnerships and realized the maximum win-set alliance.

China's progress and achievements in global climate governance have just benefited from this route. The successful transformation has strengthened China's confidence to take an active part in global governance and also exhibited China's reform and opening-up results to the world. As to the climate change issue, the developed countries try to evade their due responsibilities and shift these responsibilities to the emerging developing countries. Without financial and technical aids from the developed countries, the emerging economies at the industrialization stage have inherited the carbon-intensive development pattern prevailing in Europe and U.S. more than one century ago and aggravated the momentum of climate deterioration. After maintaining the reform and opening up for more than four decades and reaching certain level of economic development, China has realized the problems hiding behind the carbon-intensive pattern of the developed countries and selected a low carbon development route. By adhering to this strategic choice plus the two-level strategic vision and the opening-up thought of multi-stakeholder cooperation, China will make more demonstrative contributions to global climate governance.

2 Outlook

The U.S. President Trump announced the U.S. withdrawal from the Paris Agreement in the White House on June 1, 2017, Washington Time. In that November, the COP23 was held in Bonn, Germany. The US Federal Government sent a delegation with only 7 members and didn't open an U.S. Pavilion on the venue. The delegation didn't hold any news press and the government representatives attended only one side event except the formal negotiations. However, that side event involved supporting fossil energy and was interrupted by the protestation of NGOs.

When different stakeholders felt irrigated and disappointed with the climate policy of the US Federal Government, the representatives from the state governments, cities, business, research institutes and NGOs of U.S. assembled in a large tent at the exit of the main venue of the Bonn negotiation, and organized various activities. Their slogan is that "We Are Still In". It's a kind of symbol that a new era is coming.

2.1 From Top to Bottom: Way of Reform of Regime Complex

Since the 1990s, the top designing of the global climate governance has taken shape and made a top-down institutional arrangement with the core coordination of the UNFCCC. Some climate-related institutional arrangements overlap with one another, thereby forming a 'regime complex' (Raustiala & Victor, 2004) (see Fig. 2).

From this governance structure chart published by the IPCC, we can clearly see that the "regime complex" comprises the UNFCCC in the central position and the distribution of international organizations, conventions/agreements, networks and action plans pertaining to climate governance. These different actors appear at the international, national/regional and local levels, and most of them work across different levels. This chart displays the current institutional arrangements and also reveals the realistic challenges faced by the top-level design, including randomness, extensive coordination and watered-down agenda.

The design is random, because it stresses the central position of the UNFCCC, but all actors involved in the design are not subject to mandatory relations with the UNFCCC.

Besides, the UNFCCC, as the major coordinator, can only play the role of generalized coordination to maintain the conceptual consistency. Under this situation, the relationships between the actors and the Convention are loose.

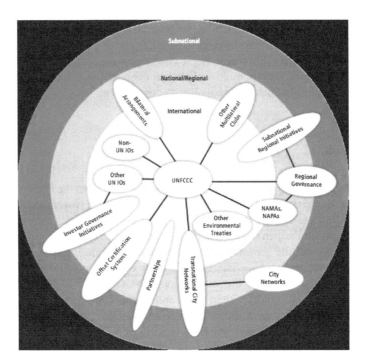

Fig. 2 Structure of global climate governance. *Source* The Fifth Assessment Report of IPCC

Besides, other multinational groups, environmental agreements and elements are all included in the governance structure, but whether climate agendas take a major position within these groups or agreements will depend on the extent of attention from the governments of the member states that changes from time to time.

Climate change response is a systematic project. The lack of the systematic coordination mechanism in the top-level design for global climate governance has made the structure a presentation of the ideal, and the lack of the institutional arrangement in reality has plunged the top-down governance pattern into a dilemma of being rigid. When facing the threat from the U.S. withdrawal from the Paris Agreement, the global climate governance has again fell into a doubt of regime failure, and the international community calls for a new leadership to reform of the global climate governance regime.

2.2 From Bottom to Top: New Drive for Local Climate Change Response

When the top-down institutional arrangement for global climate governance suffers a deadlock, let's look back to the "dual-layer, multi-dimensional" research space and observe those subtle changes that have happened in the two-level game field where China stays from the details but not presented in the governance structure at the national level.

The U.S. declaration to withdraw from the Paris Agreement has resulted in a new adjustment of the forces of game faced by China at the international level. As to the transition of leadership, France and other countries have actively voiced their opinions in the hope of joining hands with China to co-lead the climate change response. At the COP 23 in Bonn in 2017, the vast number of developing countries displayed unprecedented unity and the developed countries also exhibited great flexibility and constructiveness in the context of absence of U.S.

The surveys of China4C and YPCCCin 2017 reveal that the majorities of the publics in both China and U.S. support respective government's decision to sign the Paris Agreement and evolve towards low carbon and new energy patterns. 95% of Chinese people support Chinese Government's decision to implement the Paris Agreement, while 64% of the U.S. public oppose the U.S. decision to withdraw from the Paris Agreement (see Fig. 3).

This shows that China is now steadily working on the international level to make corresponding preparation for next step. Compared to steady progress at the international level, stakeholders at the domestic level, particularly non-state actors, have displayed unprecedented vitality.

1. **The Opportunity of the Framework Has Strengthened the Conceptual Consensus**

Framework setting means setting certain issue in an appropriate background to achieve an expected explanation or view. Framework setting doesn't intend to mislead or manipulate people's thoughts but help the public understand climate change and

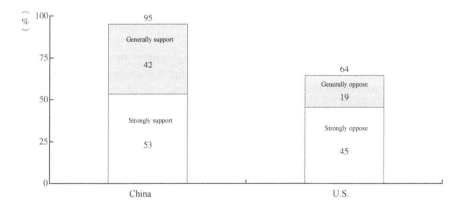

Fig. 3 Public attitudes toward withdrawal from the paris agreement in China and the U.S. *Source* A joint comparative study performed by YPCCC and China4C, published on November 10, 2017. Searched from the website of the UNFCCC

its impacts in an easier manner. As the framework can be set, climate communication actors (including government, media, NGOs, scientists, enterprises and common public) can consciously choose a framework that can trigger a resonance.

China4C's 2017 national surveyshows that if China doesn't take any actions to address climate change over next two decades, 95.1% of the respondents believe climate change will cause an increase of air pollution followed by disease epidemics. 33.4% of the respondents have selected "aggregation of air pollution" when answering the question "what climate change impact do you worry about most?". More than 70% of the respondents believe there is synergy between climate change and air pollution.

Haze and air pollution have become the most important health threats for Chinese in recent years. Actually, air pollution and climate change share the same source (emissions from fossil energy consumption), and both demand "emissions reduction" as the solution. Therefore, if we introduce haze and air pollution to climate change communication, we can solve the problem that the public think climate change is "remote and unreachable" and lacks forecast and real time effect, reinforce their awareness of climate change and make a preparation for actions.

2. **Technological Innovation Promotes Public Participation**

Climate governance needs to take more tangible actions, while the biggest challenge for climate change communication is to reshape the public's perception, attitude and behavior. The survey reveals that more than 50% of Chinese people have used the bike sharing service, more than 90% of the public support bike sharing service as a mobility solution. In the last 3 years, the Chinese government encouraged households to install rooftop solar photovoltaic panels by allowing electricity generated from solar panels to be sold to the State Grid. It is a relatively new policy which may directly induce lifestyle change of the general public. According to China4C's 2017 survey, 55.6% of the respondents said that they were aware of this policy already that the sur- plus electricity generated by solar PV can be sold to the State Grid after household/business consumption. Compared to the 2012 survey, the public could only select traditional solutions such as turning off the light, lower the air conditioner at that time. Five years later, sharing economy and technological innovation provided new possibilities for the public to take climate actions.

Besides the possibility of personal action, the public also have the strong willingness to influence people around them. The survey shows that 97.7% of the public are willing to share climate change-related information with friends and family members around them. Besides influencing people around them, the public have also paid much attention to the impacts of climate change on future generations. 98.7% of the respondents support the implementation of climate change-related education at schools (see Fig. 4).

We can forecast from these data that China will embrace a new era when "everybody is a climate communicator and climate actor".

2 Outlook

98.7% of the interviewees support the climate change-related education at schools

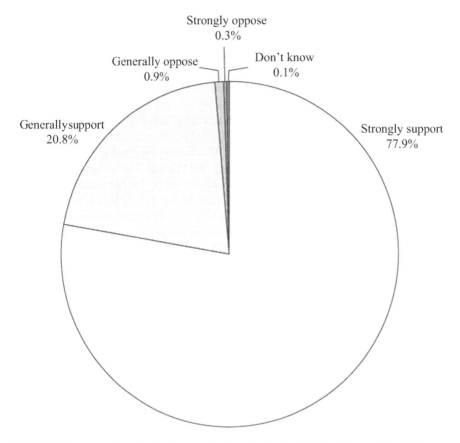

Fig. 4 Public support for climate change education in schools in China. *Source* Climate Change in the Chinese Mind 2017. China4C. (2017). Retrieved from. http://www.efchina.org/Reports-en/report-comms-20171108-enChina4C

3. Innovation Drive of NGOs and Enterprises

The latest survey has found that the public generally think the government should do more to address climate change, followed by media and NGOs. After the in-depth work for five years, the public have acknowledged the roles of the three key stakeholders of climate change communication and governance and developed stronger expectation to them.

The international NGOs working in China have actively implemented some interesting innovation while disseminating ideas, raising public awareness and advocating policies. Greenpeace and the California-based clean energy fund New Energy Nexus jointly initiated Power Lab, a clean energy incubator, which aims to identify high-quality energy innovation projects, individuals and teams in China. The incubator will provide comprehensive support, including capacity building, mentor's advice, network building, communication and other aspects, and help winners access investment, financing and commercial incubator resources.

While promoting climate governance, the NGOs have built online platforms to take an active part in new agendas of global governance. The "Belt and Road Green Development Platform" was jointly sponsored by the Natural Resources Defense Council, Oxfam, the Greenhub, WWF, and other organizations.[1] The platform aims at realizing the 2030 Sustainable Development Goals and the objectives of the Paris Agreement, and focuses on ecological environment conservation, climate change response, energy transformation, green finance, industrial cooperation and other areas involved in the "Belt and Road Initiative". Meanwhile, the platform aims to exert its advantages and advise China to implement the "Belt and Road Initiative" in the environment-friendly way and benefit the local community. The platform coordinated a series of side events under the framework of the Belt and Road Initiative in COP23 in 2017, and the themes ranged from multilateral developmental finance to climate financing and further to autonomous contributions of countries, and from renewable energy to south-south climate cooperation. These meetings were widely reported by media. In particular, the English reports of the Xinhua News Agency were quoted by the Information Office of the State Council to display the achievements NGOs have made from active involvement in the implementation of the "Belt and Road Initiative" to the international community.

Besides the NGOs, more and more local enterprises have participated in the mission of addressing climate change. For instance, Wang Shi, founder of Vanke Group, which is a famous real estate company in China, is the outstanding delegate in the field of climate from the very beginning. He participated the COP15 in 2009 and was impressed by the urgency of climate actions. Since then, he has dedicated himself to climate change response. Wang Shi promoted green transformation of his company and guided conceptual innovation among entrepreneurs. After the COP21 in 2015, Wang Shi founded C-Team, a climate alliance of local entrepreneurs. During the COP23 in Bonn in 2017, the alliance announced the "Low Carbon Initiative" on behalf of 450,000 Chinese enterprises.

As to more public to take real climate actions, Alipay initiated the "Ant Forest" campaign, which has become a phenomenon-grade innovation project and got the "Champions of the Earth" awarded by UNEP in September 2019. The action encourages Alipay users to perform low carbon behaviors such as subway travel and online payment to reduce carbon emissions. Concrete carbon emission can be used to plant trees in the game on Alipay and plant a tangible tree in reality. The latest data disclosed by the Ant Forest show that it had registered more than 500 million users, cumulatively planted 12.2 million real trees. At the same time, the Ant Forest has attempted to become an incubator for social philanthropy innovation and newly announced "the Planet Blue" initiative which will open product capacity and technological platforms to the entire society and call everybody to participate in a green future.

---

[1] For details, please see the website of the Global Green Leadership: http://www.chinagoinggreen.org.

4. Philanthropy Provides New Drive

Accompanying the growth of China's comprehensive strengths, a crowd of local entrepreneurs have grown up and participated in the philanthropy cause by setting up foundations or cooperating with charity organizations while operating enterprises. In 2016, China issued the national Charity Law, which defines related institutional arrangement. Afterwards, Internet donation, charity trust, corporate social responsibility and other aspects have taken on new looks. These latest practices have happened in the area of climate governance and become the new drive.

In September 2017, Laoniu Foundation, founded by the local entrepreneur Gensheng Niu, donated 74.38 million yuan to China Green Carbon Foundation, which will use the donation to plant more than 30,000 hectares of forest in the areas around the venue of 2022 Winter Olympic Gamesin Chongli District, Zhangjiakou City. The carbon sink forest is expected to absorb about 380,000 tons of carbon dioxide in the atmosphere during the 30-year project period. The Laoniu Carbon Sink Forest for Winter Olympic Games is a significant philanthropic project that will highlight the themes of climate change response and green development and focus on the 2022 Beijing-Zhangjiakou Winter Olympic Games.

The greater significance of the project is to bring philanthropy into the scope of public-private partnership (PPP) in practice and inject a new drive for traditional PPP mode. As a new pilot in the climate area, the project has blazed the trail for the updated PPP2.0 mode featuring "public-philanthropy-private partnership" (see Fig. 5).

More and more philanthropists have come onto the front stage of climate governance and become positive stakeholders. According to the experience of international counterparties, joint action has become the inevitable choice to maximize

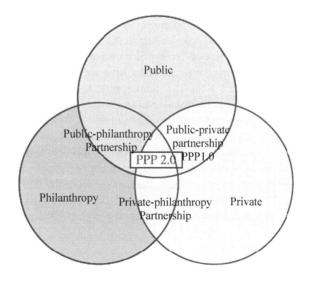

Fig. 5 PPP2.0-"Public-Philanthropy-Private Partnership". *Source* Prepared by the author

social benefits of philanthropic funds. On January 29, 2018, the China Environmental Grantmakers Alliance was inaugurated by ten local foundations, including Beijing Qiaonyu Foundation, Laoniu Foundation, Vanke Foundation, China Green Carbon Foundation and Alashan SEE Conservation.

Eight months later in San Francisco, more than ten senior delegates from Chinese philanthropies, think tanks and NGOs jointly announced the Global Climate Action Initiative in the Global Climate Action Summit to forward the bottom-up momentum from China to the world. Michael Bloomberg, former Mayor of New York and UN Special Envoy for Climate Action, formally accepted the invitation by Zhenhua Xie, Special Representative for Climate Change Affairs of China and President of Institute of Climate Change and Sustainable Development at Tsinghua University to serve as co-chair of this initiative. The Global Climate Action Summit in California not only witnessed the grand launch of the Global Climate Action Initiative, but also provided the best opportunity and stage for Chinese philanthropists in the climate change area to speak collectively for the first time on the international stage. Global Climate Action Initiative is the first global platform for cooperation and communication among politicians, entrepreneurs and philanthropists focusing on tackling the climate change launched by China. It will have a profound impact on China's role as a participant, contributor and leader in dealing with global climate change.[2]

Since the 1990s, collective decision-making has emerged as an obvious trend of global governance. With the accumulation of practical experience and reflection of earlier work, the United Nations has realized that the participation of other intergovernmental organizations, nongovernmental organizations, private sectors and entire civil society will be of important strategic significance to accomplish global sustainable development.

Chinese President Xi Jinping stressed in the report at the 19th Party Congress that China should adhere to common governance by all and establish an environmental governance system with government as the leading force, enterprises as actors, and social organizations and public as common participant. Guided by new governance concepts, Chinese stakeholders at the domestic level have actively taken actions (see Fig. 6). In the time of "everyone is climate communicator", technological innovation will inspire more actions, and China will march towards a green and low carbon future. China's actions at the domestic level will positively promote more positive actions at the international level and further join hands with all countries worldwide to address climate change together and build a "shared community of human destiny".

To realize genuine global governance and cooperation, we should also have a bottom-up roadmap besides a top-down decision-making mode. The top-down mode and the bottom-up mode will evolve in two single directions at the early stage of global climate governance. We can see from the researches in this book that while the United Nations is promoting global climate governance in a top-down manner, the bottom-up forces have emerged within countries.

---

[2]For details, please visit: http://www.cgpi.org.cn/content/details42_5606.html.

2 Outlook   143

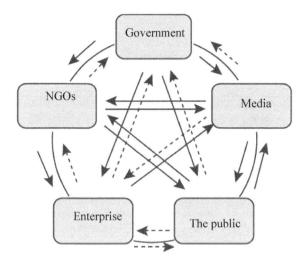

Fig. 6 Ideal mode for climate change communication and governance. *Source* Prepared by the author

Fig. 7 "Convective Zone" of global climate governance. *Source* Prepared by the author

Looking ahead to the future, the top-down mode and the bottom-up mode will meet each other halfway, forming a governance pattern like "convective zone" with self-circulation (Fig. 7). China is now changing fast every day. The experiences and lessons China learnt from the climate governance has been introduced and referred to more broader governance issues to some extent. The success of co-governance in the climate field demonstrates a deserved China's route for the future to play more constructive role in the world.

Appendix 1
Climate Change in the Chinese Mind 2012

Remarks

Climate change is a serious challenge that humans face in the 21st century. In recent years, climate change impacts have increasingly emerged along with the frequent occurrence of extreme climate events across the globe, including intense heat, drought and flood. China has paid great attention to climate change for a long time. As we know, China is a developing country with scarce resources per capita, fragile ecological environment, and frequent natural disasters. Meanwhile, the country is under the most serious climate change impacts but has relatively weak capacity to respond to climate change.

The Chinese government deems the response of actively acting on climate change as a significant opportunity to promote the transformation of economic development mode and the adjustment of economic structure by adopting a series of critical climate mitigation and adaptation policy measures. In 2011, the National People's Congress examined and passed the *Outline of the Twelfth Five-Year Plan for National Economic and Social Development*, which put forward that China should adhere to comprehensive, coordinated, and sustainable development, accelerate the transformation of economic development mode, and further take actively tackling climate change as well as advancing green and low-carbon development as a policy of significance.

The formulation and implementation of sustainable development strategies is closely bound up with everyone, thus requiring all the people and all sectors of society to participate and respond. The public need to take part in the cause of addressing climate change, because the problem cannot be solved unless everyone keeps a watchful eye on climate change issues and starts to take action from himself bit by bit. Public attitudes and demands must be identified to better encourage their participation. The survey covering 332 prefecture-level units and over 4,000 samples was conducted by the China Center for Climate Change Communication, which objectively and impartially revealed the public awareness, views and advice regarding the climate change based on a public awareness investigation in Mainland China from a third-party perspective; in addition, the questionnaire showed that the survey has understood and covered the complicated and comprehensive topic of climate change in an inclusive method, ranging from scientific, economic, social to many other dimensions.

The meaningful survey provides important referential value for the relevant parties to identify the current state of public climate change awareness and to hatch pertinent policy measures. It is hoped that the China Center for Climate Change Communication will further carry out such work and attract more people to join so as to make continuous contributions to answer and tackle climate change in partnership with all sectors of society.

XIE Zhenhua
Deputy Director of the National Development and Reform Commission

Appendix 1: Climate Change in the Chinese Mind 2012

The survey was conducted in 2012 by the China Center for Climate Change Communication (China4C), which was jointly established by the Research Center of Journalism and Social Development of Renmin University of China and Oxfam Hong Kong in April 2010. China4C is committed to the research on the climate communication theory and practice in international climate negotiations, and on China's climate change strategies in policy formulation and implementation.

The data collection and statistical work for the survey was completed by the School of Statistics at Renmin University of China, which is equipped with the National Key Discipline of Applied Statistics and known as a key research base established by both the Ministry of Education and the National Bureau of Statistics. In particular, the dean of the school, Yanyun Zhao, and his team took charge of data collection and statistics for the survey.

Survey Lead

Baowei Zheng
Director of the China Center for Climate Change Communication

Principal Investigators

Binbin Wang	Yujie Li	Mei Lyu
Yang Song	Gangcun Li	Yuanyuan Ren
Yanyun Zhao	Yan Jiang	Nan Xiao

Consultants

Qizheng Zhao Director of the Consultative Committee of China4C, Director of the External Affairs Committee of CPPCC, Dean of the School of Journalism of Renmin University of China

Zhenhua Xie Director of the Consultative Committee of China4C, Deputy Director of the National Development and Reform Commission

Yulu Chen	Director of the Consultative Committee of China4C, President of Renmin University of China
Shengrong Ma	Former Deputy Director and Deputy Chief Editor of the Xinhua News Agency
Wei Su	Chief of the Department of Climate Change of the China National Development and Reform Commission (NDRC)
Fulin Chi	President of China Institute for Reform and Development
Jiankun He	Deputy Director of National Climate Change Expert Committee
	Former Managing Vice President of Tsinghua University
Bugao Wen	Press Office Director of the NDRC
Ji Zou	Deputy Director of National Center for Climate Change Strategy and International Cooperation(NCSC)
Xuebing Sun	Policy Advocacy Director of Oxfam Hong Kong
Anthony Leiserowitz	Director of the Yale Project on Climate Change Communication (YPCCC)
Alex Kirby	Former Senior Environmental Journalist of British Broadcasting Corporation
Dennis Pamlin	Project Leader of the "21st Century Frontiers"
	Policy Advisor of United Nations Global Compact

Table of Contents

Introduction
Survey Method
Executive Summary
Main Content

    A. Climate Change Beliefs
    B. Climate Change Impacts
    C. Responding to Climate Change
    D. Support for Climate Policies
    E. Enforcement of Climate Actions
    F. Climate Change Communication

Appendix: Sample Demographics

Introduction

China is one of the countries which suffer the greatest adverse impacts from climate change. Due to relatively unfavorable climate conditions and frequent meteorological calamities, it has met with globally rare disaster-related dilemmas, including

Appendix 1: Climate Change in the Chinese Mind 2012

extensive areas, excessive types, serious conditions, and a large population affected. Climate change has posed increasingly obvious hazards to China's natural ecosystem, as well as its economic and social development in recent years.

Meanwhile, as the world's largest developing country, China has a large population with limited energy resources. With unbalanced development, the country is still undertaking its historical mission of complete industrialization and urbanization. To this day, there are still more than 100 million poverty-stricken people living in China, which entails arduous work to develop the economy, eliminate poverty, and improve people's livelihoods.

Notably, both of China's per capita and historical carbon emissions are not on the same scale as in developed countries. However, China has become the world's largest emitter of greenhouse gases and energy consumer in rapid economic development over a decade.

In this context, notwithstanding a developing country with the world's largest population that is quite fragile to climate change, China still faces much pressure in the United Nations climate change negotiations as the top emitter. Thus, it's not only challenging but also imperative for China to actively tackle climate change, promote energy conservation and emission reduction, as well as develop green economy domestically.

With profound understanding of the complicacy and extensive impacts of climate change, the Chinese government has attached great importance to the issue and incorporated it into China's mid-and-long term economic and social development planning as a major topic that influences the overall development.

Tackling climate change requires not only the active guidance and efforts of the government but also the proactive participation in low-carbon activities from the public. The relevant policies and measures can only be put into practice when everyone lives, consumes, and acts in a low-carbon and energy-saving way.

To better understand and grasp the public awareness of climate change and the relevant topics, the China Center for Climate Change Communication conducted the nationwide survey from July to September, 2012. The survey aimed to investigate and analyze the public perception of climate change issues, climate change impacts, the response to climate change, the support for climate policies, the enforcement of climate change countermeasures, and the evaluation on climate change communication.

By collecting the aforementioned information, we hope to provide reliable data support to raise public awareness of climate change, to improve public adaptive capacity for climate change and to motivate the public to act in response to climate change. This survey could also serve as a basis for the government and relevant stakeholders to make decisions and formulate measures to construct a resource-saving and environment-friendly society, to improve the capacity to mitigate and adapt to climate change, and to protect global ecology, etc.

In consideration of the complicacy in both climate change issues and China's national conditions, the survey might fail to comprehensively reflect the public understanding of climate change and diverse attitudes towards the issue, but we do hope it could help relevant governmental departments, the academia, and other institutions to gain multi-dimensional knowledge of the current public perception of climate change with some valuable information.

The Center conducted the survey following three projects, i.e. "Research on the Roles and Influences of Governments, Media and NGOs in the Post-Copenhagen Era", "To Cancun—Acts for Post-Copenhagen Climate Change Communication", and "Poverty, Climate Change, and Public Communication". The survey has deepened the empirical research based on these earlier projects. We will track survey questions hereof to constantly enrich and improve China's research on climate change and climate communication, facilitating China to lay a solid foundation to respond and adapt to climate change.

Survey Method

1. Respondents: Residents aged from 18 to 70.
2. Time: July to September, 2012.
3. Scope: Mainland China (excluding Hong Kong, Macau, and Taiwan).
4. Method: Thanks to high popularity rate of fixed-line and mobile phones in Mainland China, the survey was a computer aided phone survey (CATI). Specifically, samples were drawn from 60.0% fixed-line phones and 40.0% mobile phones.
5. Number of samples: The CATI covers 4,169 respondents.
6. Sampling Plan: In light of the 332 prefecture-level administrative units (including 284 prefecture-level cities, 15 regions, 30 autonomous prefectures, and 3 leagues) and 4 municipalities directly under the central government in China, the total population was divided into 336 levels. The sample numbers were assigned to such levels in population proportion, contributing to proportional sampling. Concretely, the phone numbers of residents were drawn at random by the tail number, with the fixed-line and mobile phone numbers respectively accounting for 60.0 and 40.0%. Ultimately, 4,169 valid questionnaires were acquired.

Executive Summary

From July to September of 2012, the China Center for Climate Change Communication and the School of Statistics of Renmin University of China conducted a national survey of 4,169 Chinese adults in Mainland China, using a sample combining urban and rural residents. The survey aimed to understand public awareness, attitudes, practice, etc. in relation to climate change and the relevant topics. The survey margin of error is +/− 1.54%. Some highlights are as follows:

Appendix 1: Climate Change in the Chinese Mind 2012

A. **Climate Change Beliefs**
   - 93.4% of respondents say they know at least a little about climate change. Specifically, 28.4% say they know just a little about it, 53.7% know something, and 11.4% know a lot. 6.6% have never heard of climate change.
   - 93% of respondents think climate change is happening. During the survey, more than 90% of respondents of different age groups hold such a view.
   - 60.6% say that climate change is caused mostly by human activities, while 33.1% and 4.2% respectively say that climate change is caused mostly by natural changes in the environment and other factors. Only 2.1% suppose climate has not changed at all.
   - 77.7% of respondents say they are either very (23%) or somewhat (54.7%) worried about climate change. 14.2% are not very worried and 8.2% are not worried at all.

B. **Climate Change Impacts**
   - 61% of respondents say they have already personally experienced the effects of climate change while 39% hold the opposite view.
   - 68.4% say that people in China are already being harmed by climate change.
   - 57.7% think climate change will harm themselves and their families to a great deal or a moderate amount; 83.5% think it will harm the public in China to a great deal or a moderate amount; while 88.6% think it will impact future generations either a great deal or a moderate amount.
   - From the change climate impacts on rural and urban residents, 47.9% argue the impacts on rural residents will be greater.

C. **Responding to Climate Change**
   - 47.5% of respondents agree and 22.6% somewhat agree with the statement, "Human beings can adapt to climate change".
   - 76.3% of respondents agree on the statement, "If we do not change our behaviors, it will be hard to meet the challenges caused by climate change".
   - 87% of respondents say they are willing to spend 10% (roughly 26.6 %, the largest proportion) or 11 to 20% (roughly 26.2 %) more on climate-friendly products; 13% are reluctant to spend more on such products.

D. **Support for Climate Policies**
   - 87.7% favor mandatory requirement for enterprises to meet higher environmental standards in despite of higher costs.
   - 90.2% favor mandatory requirement for automakers to produce more climate-friendly cars in despite of higher costs.
   - 90.3% favor mandatory requirement for using green building materials and designs in despite of higher costs.
   - 74.4% favor mandatory requirements for consumers to buy renewable products in despite of higher costs.

- 91.9% favor mandatory standards for rubbish classification and recycling in despite of higher costs.
- 84.3% favor mandatory requirements for farmers to use organic fertilizers in despite of higher costs.

E. **Enforcement of Climate Actions**

- 83.6% always or often turn off lights in time when unnecessary.
- 79.3% always or often turn off electronic products (such as televisions and computers) in time when unnecessary.
- 47.7% always or often use reusable shopping bags rather than plastic bags.
- 33.9% always or often classify rubbish.
- 61.5% always or often reduce the use of disposable paper cups or tableware.
- 53.6% always or often reuse articles, instead of buying new ones.
- 44.7% always or often reduce the use of air conditioners as much as possible.
- 79.7% always or often save domestic water as much as possible.
- 69.5% always or often walk, ride a bike, or take public transportation.

F. **Climate Change Communication**

- Respondents say they have obtained information about climate change through the television (93.8%), telephone (66.1%), or the internet (65%).
- Respondents trust scientific institutes and the government the most as sources of information about climate change.
- When asked which kind of news they care about most, only 9.2% select environmental news.

Main Content

A. Climate Change Beliefs

A1. How Much Respondents Know Climate Change

In this survey, above 90% of respondents say they know climate change (93.4%) to varying degrees. Specifically, 53.7% know something, and 11.4% know a lot (Fig. 1).

Knowledge	Never heard of	A little	Some	A lot
Percentage	6.6%	28.4%	53.7%	11.4%

Fig. 1 How much respondents know climate change

Appendix 1: Climate Change in the Chinese Mind 2012

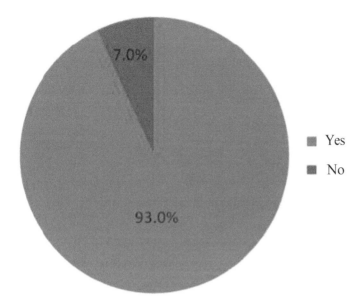

Fig. 2 Whether climate change is happening

A2. Whether Climate Change is Happening

As for the statement, "Climate change refers to the change in the average state of climate with the lapse of time, do you believe climate change is happening", 93% of respondents think climate change is happening. During the survey, more than 90% of respondents in each age group hold such a view (Fig. 2).

A3. Causes of Climate Change

With respect to the causes of climate change, 60.6% say that climate change is caused mostly by human activities, while 33.1% say that climate change is caused mostly by natural changes in the environment. 2.1% suppose climate has not changed at all (Fig. 3).

A4. How Much Respondents are Worried about Climate Change

77.7% of respondents say they are either very (23%) or somewhat (54.7%) worried about climate change. 14.1% are not very worried and 8.2% are not worried at all (Fig. 4).

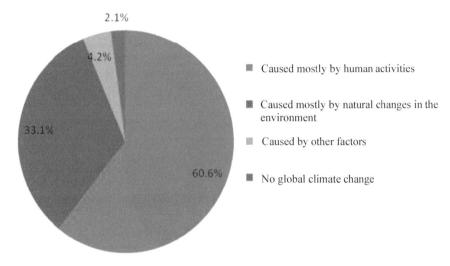

Fig. 3 Causes of climate change

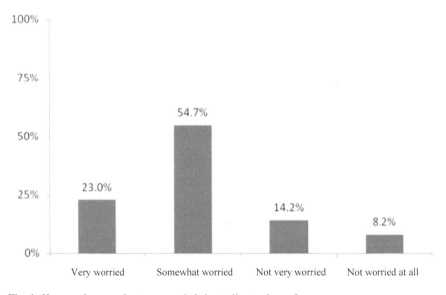

Fig. 4 How much respondents are worried about climate change?

B. Climate Change Impacts

B1. Climate Change Experience

61% of respondents say they have already personally experienced the effects of climate change, which is 20% higher than those who believe they have not yet experienced it (Fig. 5).

B2. The Time frame of the Harm Caused by Climate Change

Regarding the statement, "Do you think China will be harmed by climate change? It is already being harmed or in how many years will it be harmed? In 10, 25, 50, or 100 years, or never", 68.4% say that people in China are already being harmed by climate change (Fig. 6).

B3. Judgment on Who Are Affected by Climate Change

In general, respondents believe that climate change will have a great deal or a moderate amount of impacts on themselves, their families, the public, and future generations. In particular, respondents think climate change are most impactful to future generations, which are followed by the public, themselves, and families (Fig. 7).

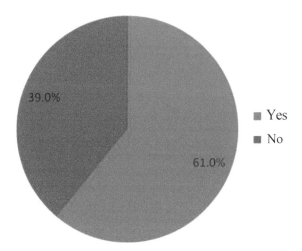

Fig. 5 Whether respondents have experienced climate change

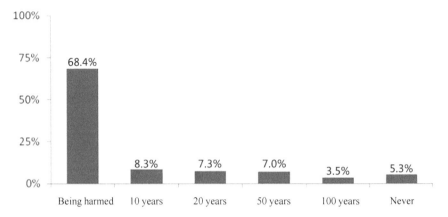

Fig. 6 Whether China will be Harmed by Climate Change

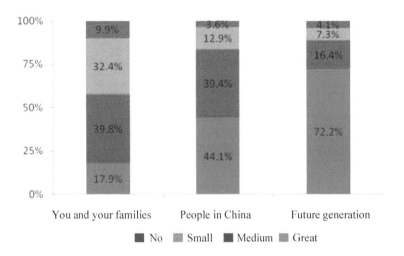

Fig. 7 Climate change impacts on different groups

B4. Climate Change Evidence

For possible impacts caused by climate change, over 60.0% respondents think 'droughts and water scarcity (1), flood (2), more diseases (3), extinction of plants and animals (4), and famine and food scarcity (5) will increase to varying degrees "in the next two decades in China, if without any climate change countermeasures". However, they have relatively weakest crisis awareness of potential famine and food scarcity (5) (Fig 8).

	1. Drought and water scarcity	2. Flood	3. More	4. Extinction of plants and animals	5. Famine and food scarcity
Increase a lot	58.8%	47.8%	46.7%	40.1%	26.6%
Increase to some extent	31.1%	37.4%	41.8%	40.6%	38.0%
Reduce to some extent	3.7%	4.8%	2.7%	8.2%	8.5%
Reduce a lot	2.3%	2.8%	2.1%	3.8%	3.5%
No change	4.2%	7.1%	6.7%	7.4%	23.5%
Total	100.0%	100.0%	100.0%	100.0%	100.0%

Fig. 8 Climate change impacts on different phenomena

B5. Extreme Weather Event (Drought) Caused by Climate Change

For the statement, "If a severe drought of more than one year hits the region where you reside, what impact does it have on your food supply (1), drinking water supply (2), household income (3), families' health (4), and crops (6)?", Respondents think drought has huge impact on all the above elements. However, people in general worry less about climate change impacts on housing security (5) as most chose it has "some impacts", while the percentage of choosing that it has "great impact" is the smallest (Fig 9).

C. Responding to Climate Change

C1. People's Confidence in Responding to Climate Change

47.5% of respondents agree and 22.6% somewhat agree on the statement, "Human beings can adapt to climate change". 76.3% of respondents agree and 13.6% somewhat agree with the statement, "If we do not change our behaviors, it will be hard to meet the challenges arising from climate change". 44.8% of respondents agree and 16.2% somewhat agree with the statement, "Individual act can play a role in tackling climate change". 88.2% of respondents agree and 9.9% somewhat agree with the statement, "The government should pay high attention to climate change" (Fig. 10).

	1. Food supply	2. Drinking water supply	3. Household income	4. Families' health	5. Housing security	6. Crops
Great impact	53.0%	54.4%	43.0%	42.0%	19.6%	74.4%
Some impact	30.7%	25.2%	33.3%	39.7%	29.3%	17.8%
Small impact	13.2%	13.8%	16.5%	12.9%	25.3%	4.9%
No impact	3.2%	6.6%	7.1%	5.4%	25.8%	2.9%
Total	100%	100%	100%	100%	100%	100%

Fig. 9 The extent to which a drought will impact in the view of respondents

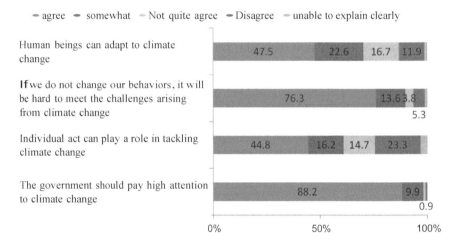

Fig. 10 Perception of response to climate change

C2. Willingness to Spend More to Combat Climate Change

87.0% of respondents say they are willing to spend 10.0% (roughly 26.6 %, the largest proportion) or 10.0 to 20.0% (roughly 26.2 %) more on climate-friendly products; 17.2 and 17.0% are willing to spend 21.0 to 30.0% and above 30.0% more; 13.0% are reluctant to spend more to buy such products. Thus, it can be seen many respondents are willing to spend 10.0–30.0% more and they make up for nearly 70.0% of the whole willing respondents (Fig. 11).

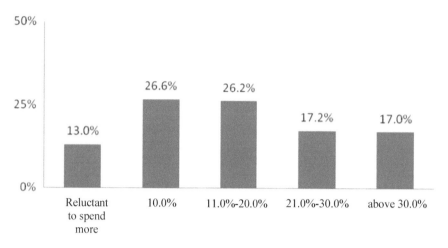

Fig. 11 Proportion of the increasing costs accepted by respondents to buy climate-friendly products

C3. Leading Roles in Responding to Climate Change

When asked about the leading roles in responding to climate change, 68.1% choose the government; around 15.9% choose the public; no more than 10% believe the media, enterprises, and NGOs should take that role.

As for the secondary role in responding to climate change, the top two picks are the media and the public, representing 28.0 and 25.5%, respectively.

With the analysis involving both leading and secondary roles, it is found that 88.9% deem the government as the agency liable for tackling climate change, followed by the public, media, enterprises/ commercial organizations, and NGOs (Fig. 12).

D. Support for Climate Policies

D1. Attitude Towards the Climate Policies

Most of respondents support all the climate policies, and a majority strongly favor them in despite of more costs. Among these responses, the highest proportion (54.4 %) of "strongly favor" were for "mandatory standards for rubbish classification and recycling in despite of higher production costs", while the lowest proportion (26.1 %) falls under "mandatory standards for consumers to buy renewable products in despite of higher consumer costs" (Fig 13).

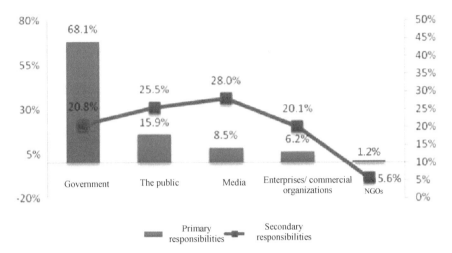

Fig. 12 Leading and secondary roles on climate change

Will you favor any of the following measures taken by the government in despite of more costs?	Strongly oppose	somewhat disapprove	somewhat approve	strongly approve
Requiring enterprises to meet higher environmental standards, even if this raises production costs	2.9%	8.1%	42.9%	44.8%
Requiring automakers to produce more environmentally friendly cars, even if this raises production costs	2.3%	6.2%	39.1%	51.1%
Using green building materials and designs, even if this raises construction costs	2.5%	5.6%	43.7%	46.6%
Requiring consumers to buy renewable products, even if this raises consumer living expenses	7.0%	16.6%	48.3%	26.1%
Requiring waste classification and new recycling standards, even if this raises costs	2.3%	4.9%	37.5%	54.4%
Requiring farmers to use organic fertilizers, even if this raises food and produce prices	4.4%	9.9%	41.1%	43.2%

Fig. 13 Comparison of recognition of governmental measures

Frequency	Always	Often	Sometimes	Seldom
Turn off lights in time when not in use	33.1%	50.5%	10.2%	3.6%
Turn off electronic products (such as televisions and computers) in time when not in use	37.1%	42.2%	12.7%	5.6%
Reuse items, instead of buying new ones when possible	19.4%	34.2%	30.2%	11.6%
Use reusable shopping bags rather than plastic bags	18.2%	29.5%	27.5%	16.1%
Minimizes use of disposable paper cups or tableware	29.6%	31.9%	20.5%	12.2%
Minimizes use of air conditioners as much as possible	18.3%	26.4%	16.5%	11.1%
Minimizes household water use (i.e. for laundry and hygiene) as much as possible	31.6%	48.1%	12.8%	5.3%
Buy local food	22.9%	49.4%	16.0%	8.8%
Classifies waste	12.1%	21.8%	18.7%	17.0%
Walking on foot, ride riding bikes, or take public transportation as much as possible	30.1%	39.4%	16.5%	11.2%

Fig. 14 Frequency of climate change countermeasures

E. Enforcement of Climate Actions

E1. Higher Rate of Implementation Among Respondents

After analyzing the frequency of implementation of different countermeasures for reducing the effects of climate change, we found that respondents implemented these countermeasures at a relatively high rate. 37.1% "always" turn off electronic products when not in use, and 50.5% "often" turn off lights when not in use. 37.1% "always" turn off electronic products when not in use, and 50.5% "often" turn off lights when not in use. However, the frequency of waste classification was relatively low. Only 12.1 % of respondents "always" classify waste and only 21.8% "often" classify waste, together totaling only one third of total respondents (Fig. 14).

F. Climate Change Communication

F1. Channels of Climate Change Information

A vast majority of the respondents were able to access to information about climate change from a variety of channels, among which the top 3 most commonly used were

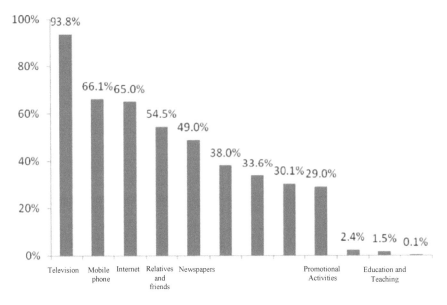

Fig. 15 Channels of climate change information

television (93.8%), mobile phone (66.1%), and the Internet (65.0%), with access to all three exceeding 60%. Relatives and friends were also a common source of information, accounting for over 50% of responses. Traditional media such as newspapers, portable media, broadcasting to magazines was also identified as an information source, but fell significantly behind modern ones in terms of influence. A minority of respondents also gain information about climate change through textbooks, school education or by personal observation and experiences (Fig. 15).

F2. The Degree of Credibility of Different Information Sources

Respondents believe the information published by scientific institutes and the government the most, followed by news media, families, and friends. They do not quite trust NGOs and enterprises (Fig. 16).

Information sources	Mean value	Standard deviation
Government	3.2	0.7
NGOs	2.2	0.9
Scientific institutes	3.3	0.7
News media	3.0	0.7
Families and friends	2.7	0.8
Enterprises	2.2	0.8

Fig. 16 Comparison of the degree of credibility among information sources

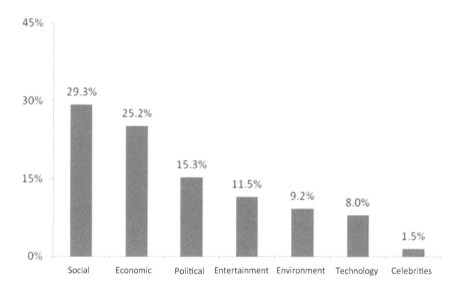

Fig. 17 News that respondents care most

F3. How Much Attention Respondents Pay to Various News Contents

In general, among the news that respondents care most, social news attracts the most attention (29.3 %), while only 9.2% respondents select environmental news as the one they care most, which demonstrates that respondents generally pay little attention to environmental news (Fig. 17).

Appendix: Sample Demographics

(a) Gender

	Frequency	Percent
Male	2406	57.7
Female	1763	42.3
Total	4169	100.0
After weighting		
	Frequency	Percent
Male	499721230	50.8
Female	483752180	49.2
Total	983473410	100.0

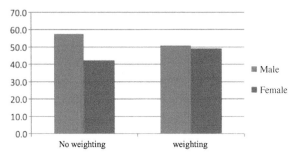

(b) Resident for the Past Year

	Frequency	Percent
Urban	2678	64.4
Rural	1480	35.6
Total	4158	100.0
After weighting		
	Frequency	Percent
Urban	595469483	60.7
Rural	385260927	39.3
Total	980730410	100.0

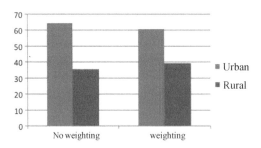

Appendix 1: Climate Change in the Chinese Mind 2012

(c) Place of Domicile

	Frequency	Percent
Urban	1898	45.6
Rural	2263	54.4
Total	4161	100.0

After weighting

	Frequency	Percent
Urban	444093372	45.2
Rural	538798942	54.8
Total	982892314	100.0

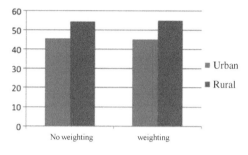

(d) Age

	Frequency	Percent
Age 18 to 24	1113	26.7
Age 25 to 34	1244	29.8
Age 35 to 44	857	20.6
Age 45 to 54	549	13.2
Age 55 to 64	297	7.1
Age 65 and above	101	2.4
Refuse to answer	4	.1
Not sure	4	.1
Total	4169	100.0

After weighting

	Frequency	Percent
Age 18 to 24	169421209	17.2
Age 25 to 34	197754675	20.1
Age 35 to 44	242694545	24.7
Age 45 to 54	183838419	18.7
Age 55 to 64	139979756	14.2
Age 65 and above	47978767	4.9
Refuse to answer	1057452	.1
Not sure	748587	.1
Total	983473410	100.0

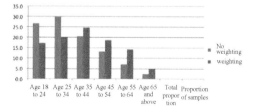

(e) Education Background

	Frequency	Percent
Elementary and below	316	7.6
Junior high	960	23.0
Senior high	982	23.6
Technical secondary	340	8.2
Two-year college	693	16.6
Undergraduate	811	19.5
Graduate and above	67	1.6
Total	4169	100.0

After weighting:

	Frequency	Percent
Elementary and below	34184911	3.5
Junior high	226887791	23.1
Senior high	448868590	45.6
Technical secondary	158240075	16.1
Two-year college	66719659	6.8
Undergraduate	44458733	4.5
Graduate and above	4113651	.4
Total	983473410	100.0

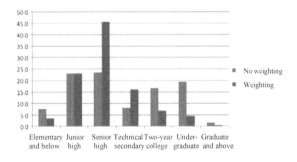

Appendix 1: Climate Change in the Chinese Mind 2012

Appendix 2
Climate Change in the Chinese Mind 2017

Remarks

The China Center for Climate Change Communication conducted the second national public awareness survey on climate change after five years. The survey report shows high awareness of climate change among the Chinese public. That respondents strongly support the government's relevant policies, and particularly that over 90 percent of respondents support the implementation of the Paris Agreement are the greatest encouragement and approval to China's efforts of addressing climate change.

I expected the data and findings from this survey would provide meaningful referential information for all sectors of society. It is hoped that the China Center for Climate Change Communication will further carry out such significant and precious work, providing scientific data for us to keep delivering "China's solutions" embedded with Chinese wisdom to the world.

XIE Zhenhua
China's Special Representative for Climate Change
October 31st, 201

The survey was designed and conducted by the China Center for Climate Change Communication(China4C) in 2017. The China4C, established in April 2010, is the first think tank among all developing countries focusing on the research about the climate change communication theory and practice, as well as research on the strategic communication analysis in China's climate change policy making and implementation.

The data collection and statistical work for the survey was completed by Survey and Statistics Institute of Communication University of China (SSI). SSI is the first university affiliated institution with domestic and foreign-related social investigation permit in People's Republic of China.

The survey was funded by the Energy Foundation. The report does not represent Energy Foundation's views.

Cite as: Wang, BB., Sheng, YT., Ding M., Lyv M., Xing JL., Zhou QN. (2017). *Climate change in the Chinese mind: 2017*. Beijing, CT: China Center for Climate Change Communication.

Survey Lead

Binbin Wang, Ph.D.

Co-founder of the China Center for Climate Change Communication

Post-doctoral Research Fellow at the School of International Affairs, Peking University

(86)138-1037-7810, http://binbinwang@pku.edu.cn

Principal Investigators

Binbin Wang, Mei Lyv, Jingli Xing, Qinnan (Sharon) Zhou

Mai Ding, Yating Shen

Consultants

Baowei Zheng	Director of the China Center for Climate Change Communication
Anthony Leiserowitz	Director of the Yale Program on Climate Change Communication

Preface

The scientific cognition of climate change involves three levels. Firstly, the understanding of climate change itself, that is, what exact changes have happened in climate? On one hand, climate change is based on an incredible amount of data reflecting observational facts, the warming trend, frequent extreme whether events and so on. On the other hand, the modern climate change science, with solid and rational knowledge, can provide a theoretical basis. Secondly, climate change in contemporary age is caused by both human and natural factors. Human activities lead to the rising concentration of greenhouse gases in the air; natural factors include changes in solar activities, the absorption of CO_2 by the ocean, and the like. Climate change is attributable to the interplay of the two factors. Modern climate change research lays emphasis on human activities, as they have increasingly obvious impacts on and a closer relation with climate change. Thirdly, in terms of impacts and consequences, climate change is a double-edged sword, but its adverse impacts on humans and the Earth have been increasingly severe. Without effective countermeasures, it might trigger a critical point where a deluge of catastrophes would emerge.

Global response to climate change is definitely a trend and direction, which serves the purpose to achieve the sustainable development of the world and all human beings, as well as to lead the green, low-carbon development pathway in the globe. The trend is inevitable, as it is the fundamental requirement for a community of shared future for mankind to pursue sustainable development.

For China, responding to climate change is not only a solemn commitment to the entire world but also an inevitable course for self-development. In international negotiations and global governance, China ought to lead the correct direction in partnership with other countries and serve as a contributor, promotor, constructor, and facilitator in responding to climate change and establishing a new world order, thus realizing win-win cooperation. Following the 18th National Congress of the Communist Party of China (CPC), the 19th Party Congress clearly pointed out that we should accelerate the structural reform to promote ecological civilization, build a beautiful country, continue to take ecological civilization as an important national strategy, and further clarify the path of green, low-carbon, and circular development.

In order to realize low-carbon development, China must design and implement the low-carbon consumption pattern, in addition to a series of measures including saving energy and raising energy efficiency, reducing high-carbon energy, and developing new energy resources. Therefore, we must jointly construct low-carbon and smart cities, take low carbon as an binding assessment indicator of new-type urbanization, and enable the public to engage in low-carbon development and ecological civilization as hosts. To some extent, low-carbon communities, enterprises, villages, towns, and even families constitute the cell of a low-carbon city. It will not only directly benefit the construction of beautiful cities and villages, but also enhance the quality and the civilization level of Chinese citizens, which is of fundamental significance for the Chinese nation to stand rock-firm in the family of nations.

Because of this, it can be said that each citizen, each family is the essential driving power of promoting low-carbon development in depth, advancing low-carbon pilot work, and creating a low-carbon society. Thus, the standing point of building a low-carbon society requires that we must understand the current public awareness of climate change, guide the public to cultivate scientific perceptions of climate change, and help the public live in a low-carbon lifestyle. From this perspective, the 2017 survey on public awareness of climate change and their attitudes towards climate change communications conducted by the China Center for Climate Change

Communication is very prompt and meaningful. I feel delighted to see the China Center for Climate Change Communication, as an independent third-party institution, can renew its efforts based on the 2012 survey and present a valuable gift to the colleagues in the field of climate change, enabling us to better understand public awareness of climate change; I also expect that through efforts from all walks of life, we will be able to make low-carbon development closer to the daily life of the public, and together to realize our common dream of beautiful China.

DU Xiangwan
Honorary Director of National Climate Change Expert Committee
Academician of the Chinese Academy of Engineering
October 25th, 2017

Table of Contents

Introduction
Survey Method
Executive Summary
Main Content

    A. Climate Change Beliefs
    B. Climate Change Impacts
    C. Responding to Climate Change
    D. Support for Climate Policies
    E. Enforcement of Climate Actions
    F. Climate Change Communication

Appendix: Sample Demographics

Introduction

Public participation is vital to address climate change. Understanding public awareness and attitudes of climate change definition, impacts and related policies in time would help policymakers formulate policies in a more rational way, and provide data reference to various stakeholders such as enterprises, NGOs, and research institutions in designing and carrying out relevant work.

Public awareness survey is a common method used internationally to understand how much the public know about climate change. Yale Program on Climate Change Communication and Center for Climate Change Communication of George Mason University have conducted survey on public awareness of climate change for more than a decade in the United States. In 2008, these two institutions put forward the theory of "Six Americas", which categorizes American people's awareness of global warming into six types for the purpose of a detailed further research. In addition, Pew Research Center, The Gallup Organization, Nielsen, BBC, etc. have also conducted surveys on public awareness of climate change in many countries. Generally speaking, even if some of those surveys touched China, most of them are only restricted to certain Chinese cities or extremely limited rural areas, and therefore the sampling methodology failed to represent the overall Chinese public awareness including both rural and urban residents. In review of domestic literatures, we found that domestic scholars have also respectively conducted surveys targeting urban residents, rural residents, enterprise managers, university students, and the public in different regions, but few surveys have been carried out nationwide.

In 2012, the China Center for Climate Change Communication conducted a survey on the public awareness of 4,169 respondents in both urban and rural areas of Mainland China to get the whole picture of public awareness, attitudes, and response in respect of climate change. According to the findings, only 6.6% of the public have never heard of climate change, and the majority of them think that climate change is happening, that it is mainly caused by human activities, and that China is being harmed there from and such harm has more severe impacts on rural residents. The findings also show that Chinese citizens strongly favor policies issued by the government in response to climate change. As the first national survey on public awareness conducted by an independent third party, the survey has provided data reference for international negotiations and domestic policy-making.

The public awareness data acquired from the 2012 survey have gained high attention from relevant national and international policymakers. Mr. XIE Zhenhua, the then deputy director of the National Development and Reform Commission, pointed out that in the preface of 2012 survey report that "the public need to take part in the cause of addressing climate change, because the problem cannot be solved unless everyone keeps a watchful eye on climate change issues and starts to take action from himself and bit by bit". In addition, China's Policies and Actions for Addressing Climate Change (2012) specially introduced the survey. Meanwhile, the survey data have positive impacts at the international level. In December 2012, the survey data were also quoted by Christiana Figueres, the UNFCCC's executive secretary, during the United Nations Climate Change Conference in Doha (COP20) to affirm China's concerted efforts in responding to climate change.

Five years later, a wave of changes has taken place at both national and international levels. Domestically, the public awareness, attitudes, and behaviors in terms of climate change have evolved along with the development of social economy, politics, culture and so on. With the rapid scientific and technological innovations, constant upgrading of energy-saving and low-carbon products, and the emergence of shared products, the pattern of energy consumption by Chinese people is undergoing some changes. Internationally, the United States' decision to withdraw from the Paris Agreement has caused some uncertainties in global climate governance.

Thus, the China Center for Climate Change Communication conducted the second national survey five years later, which adopted the same methodology as the previous one. The 2017 survey still categories questions into six sections, including "public beliefs of climate change", "perception climate change impacts", "responding to climate change", "support for climate change policies", "enforcement of climate change countermeasures", and "evaluation on climate change communication". What is different from the 2012 version is that we have incorporated new questions that reflect some latest changes in the past five years. With the new survey, we hope to update and improve relevant data, as well as to keep good track of the current status of public awareness of climate change in China.

We were kindly encouraged and supported by various partners when carrying out this survey. The Energy Foundation funded this survey, and provided enormous advice and support during the whole process with the team led by the its President in Beijing Office, Professor Zou Ji, and its Director of Communications, Ms. Jing Hui. To guarantee the funding can be applied to this specific survey project, China Green Carbon Foundation established a Special Fund for Climate Change Communication for us. In addition, the team of Survey & Statistics Institute of Communication University of China, led by Professor Ding Mai finished data collection work efficiently. At different stages of the survey project, ranging from questionnaire design to data cleaning, we gained valuable advice from various relevant stakeholders, including Yale Program on Climate Change Communication, the United Nations system in China, Embassies of Switzerland and other countries in China, the Department of Climate Change at the China National Development and Reform Commission, Center for Environmental Education and Communications of Ministry of Environmental Protection, China National Center for Climate Change Strategy and International Cooperation, China National Climate Center, Peking University, Tsinghua University, Chinese Academy of Engineering, Chinese Academy of Sciences, Chinese Academy of Social Sciences, China News Service, Weather China, Asian Development Bank, Hong Kong and Shanghai Banking Corporation, China Energy Construction Investment Corporation, Pricewaterhouse Coopers, World Wildlife Fund, Natural Resources Defense Council, Greenpeace, Oxfam, SEE Foundation, Paradise Foundation, Pear Video, "Wind Energy" Magazine, Innovative Green Development Program, Greenovation Hub, China Dialogue, and Mobike, etc.

The China Center for Climate Change Communication hereby extends heartfelt gratitude to all that have supported us!

The participation by multiple partners during the whole process of the survey enabled us to hear diverse voices, which inspired us to think in depth about the value, methodology, and possible applications of the survey. There are still a lot of meaningful things awaiting us in the field of public participation in addressing climate change. Hope to work with all of you to push more exciting changes to happen. Let's work together!

Binbin Wang
Co-Founder
China Center for Climate Change Communication

Appendix 2: Climate Change in the Chinese Mind 2017

Survey Method

1. Respondents: Residents aged from 18 to 70
2. Time: August to October, 2017
2. Scope: Mainland China (excluding Hong Kong, Macau, and Taiwan)
3. Method: Thanks to high popularity rate of fixed-line and mobile phones in Mainland China, the survey was a computer aided phone survey (CATI). Specifically, samples were drawn from 15.4% fixed-line phones and 84.6% mobile phones.
4. Number of samples: The CATI covers 4,025 respondents.
5. Sampling Plan: In light of the 332 prefecture-level administrative units (including 291 prefecture-level cities, 8 regions, 30 autonomous prefectures, and 3 leagues) and 4 municipalities directly under the central government in China, the total population was divided into 336 levels. The sample numbers were assigned to such levels in population proportion, contributing to proportional sampling. Besides, the proportion of age groups, gender groups, residencies (rural or urban areas), and the ownership of landlines and mobile phones are considered to guarantee the samples to be representative. Concretely, the phone numbers of residents were drawn at random by the tail number, the sampling of landline telephone respondents followed random selection as well.

Executive Summary

A. Climate Change Beliefs

- 2,834 out of 4,250 respondents share the first thing comes to their minds when hearing "climate change" in either a word or a phrase. Analyzing the frequency of these words/phrases with WordArt, we found that the most frequently mentioned words is "hot (mentioned 225 times)", followed by "haze (mentioned 179 times)" and "global warming (mentioned 170 times)".
- 80% of the 2,834 respondents who give a word/phrase as the first comes to their minds when hearing "climate change" rated their words/phrases as "negative".
- 92.7% of respondents say they know at least a little about climate change. Over half (57.2%) say they know "just a little about it", nearly one in three (31.5%) say they know "something about it", and only 4% say they know "a lot" about climate change, while 7.1% say they have "never heard of it".
- 94.4% of respondents think climate change is happening. By contrast, only 5.3% think climate change is not happening.
- 66.0% of respondents understand that climate change is caused "mostly by human activities", while 11.1% say it is due to "natural changes in the environment". And 19.3% of respondents think it is caused by both reasons. Besides, 1.7% think that climate change "is not happening".

- 79.8% of respondents say they are "very" (16.3%), or "somewhat" (63.5%) worried about climate change. 16.2% and 3.9% of respondents say they are "not very" or "not at all" worried about it, respectively.

B. **Climate Change Impacts**

- 75.2% of respondents have already personally experienced impacts of climate change while 24.6% hold the opposite view.
- 31.1% think climate change will harm themselves and their families to a great deal or a moderate amount; 51.4% think it will harm the public in China to a great deal or a moderate amount; 78% think it will impact future generations either a great deal or a moderate amount, while 71.7% think it will impact plant and animal species either a great deal or a moderate amount.
- 95.1% of respondents think climate change will cause an increase in occurrence of air pollution, followed by disease epidemics (91.3%), droughts and water shortages (89.8%), floods (88.2%), glaciers melting (88.0%), extinctions of plant and animal species (83.4%) and famines and food shortages (73.4%) in the next two decades in China, if without any climate change countermeasures.
- 33.4% of respondents worry about air pollution the most. Others are most worried about disease (29.0%), droughts (10.9%), floods (8.6%) and glaciers melting (6.8%).
- 72.6% of respondents think climate change and the air pollution are interrelated with each other. Besides, 14.3% think that climate change leads to air pollution and 12.8% think air pollution leads to the climate change.

C. **Responding to Climate Change**

- 47.8% of respondents believe that climate change mitigation is more important than climate change adaptation as countermeasures in addressing climate crisis. 45.3% of respondents think that mitigation is as important as and adaption. Only 6.7% think adaption is more important.
- When asked about the leading roles in responding to climate change, among the government, environmental NGOs, enterprises/business organizations, the public and the media, most respondents believe the government should shoulder relatively more responsibilities, followed by "the Media" and "environmental NGOs".
- When asked about which fields that the central government should pay attention to, among "air pollution, water pollution, climate change, ecosystem protection, economy development, education, terrorism and health care", over 70% of respondents think all these areas should be given particular attention from the central government. Averagely, respondents say the issue of air pollution is the most important, followed by water pollution, ecosystem protection, healthcare and climate change.
- Among the aforementioned areas that are of high public attention, 24.3% of respondents think air pollution is the most important, followed by ecological protection (18.0%) and health (17.2%), 8.8% of respondents believe that the

Appendix 2: Climate Change in the Chinese Mind 2017

most important issue of climate change, which is even more critical than economic development and anti-terrorism.

D. **Climate Change Policies**
   - In 2015, China signed an international agreement in Paris with 195 other countries. 96.3% of respondents are either "somewhat support" or "strongly support" China's participation in Paris Agreement and among them, 59.3% say they "strongly support" it.
   - 94% of respondents say they support China's decision to stay in the Paris Agreement to limit the pollution that causes climate change. 52.5% say they strongly support it.
   - 96.8% of respondents support China's effort to promote the international cooperation on climate change, of which over half (54.7%) say they "strongly support" it.
   - 96.9% of respondents support government's efforts of the total quantity control on China's greenhouse gas emissions and 64.5% are strongly supportive of it.
   - Each policy to mitigate climate change or reduce emissions is "somewhat" supported or "strongly" supported by around 90% of respondents.
   - Each policy to adapt to climate change is "somewhat" supported or "strongly" supported by over 90% respondents.
   - 98.7% of respondents support the statement that schools should teach students about the causes, consequences, and potential solutions to climate change.

E. **Enforcement of Climate Actions**
   - When asked if they would be willing to pay more for the climate-friendly products, 73.7% of respondents give the affirmative answer.
   - When asked if they would like to pay to offset their personal emissions completely (If offsetting of the personal emissions cost RMB 200 yuan per year), 27.5% of respondents say yes.
   - 46.7% of respondents have used shared bikes.
   - 92.6% of respondent support using shared bike as a way of travel.
   - 55.6% of respondents have heard that besides household consumption, electricity generated from solar photovoltaic panels can be sold to the State Grid.

F. **Climate Change Communication**
   - Respondents say they have obtained information about climate change through three major information channels, which are television (83.6%), WeChat (79.4%), and friends and family (68.1%).
   - 94% of respondents have strong desire of learning more about climate change. Specifically, most respondents would like to learn more about "climate change impacts".
   - Respondents trust the central government the most as source of information about climate change.

- When asked which kind of news they care most about, 12.3% select environmental news.
- 97.7% of respondents are willing to share climate change information with their families and friends.

Main Content

A. Climate Change Beliefs

A1. When You Think of "Climate Change", What is the First Word or Phrase that Comes to Your Mind? (open-ended)

2,834 out of 4,025 respondents share the first thing comes to their minds when hearing "climate change" in either a word or a phrase. The most frequently mentioned word is "hot (mentioned 225 times)", followed by "haze (mentioned 179 times)" and "global warming (mentioned 170 times)". WordArt shows all words/phrases mentioned by respondents as the figure shows below. Larger word/phrase size signifies higher frequency (Fig. 18).

A2. 80.4% of Respondents Think the First Word/Phrase that Comes to Their Minds is Negative When Hearing "Climate Change"

Among all 2,834 effective responses, 80% think the first word/phrase that comes to their minds when hearing "climate change" is "negative". Specifically, 29.5% say it is " −3 extremely negative", 26.8% rated it as "−2", and 24.1% rated it as " − 1very negative". Only 19.5% say their word/phrase has positive meaning. Clearly, the general impression of climate change among respondents is negative (Fig. 19).

A3. 92.7% of Respondents Say they Know About Climate Change

92.7% of respondents say they know about climate change to varying degrees. 57.2% say they know "just a little about it", 31.5% say they know "something about it", and only 4% say they know "a lot" about climate change (Fig. 20).

A4. 94.4% of Respondents Think Climate Change is Happening

94.4% of respondents think climate change is happening, while only 5.3% think climate change is not happening (Fig. 21).

Appendix 2: Climate Change in the Chinese Mind 2017

When you think of "climate change", what is the first word or phrase that comes to your mind?

Fig. 18 When you think of "climate change", what is the first word or phrase that comes to your mind? (Open-Ended) (n = 2834)

80.4% of respondents think the first word/phrase that comes to their minds is negative when hearing "climate change"

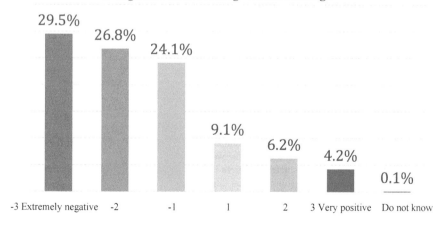

Fig. 19 Please help us to understand what that word or phrase means to you. You said: [INSERT TEXT RESPONSE] On a scale from −3 (very bad) to +3 (very good), do you think that this is a bad thing or a good thing?" (n = 2834)

92.7% of respondents say they know about climate change

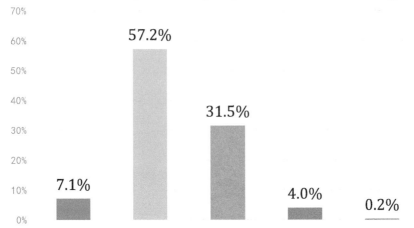

Fig. 20 How much do you know about climate change? Do you know a lot about it, something about it, just a little about it, or have you never heard of it? (n = 4025)

94.4% of respondents think climate change is happening

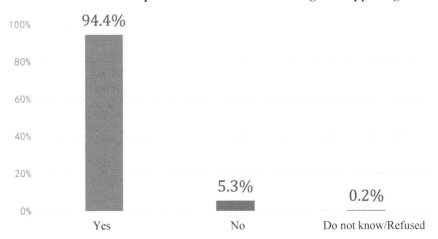

Fig. 21 Recently, you may have noticed that climate change has been getting some attention in the news. Climate change refers to the idea that the world's average temperature has been increasing over the past 150 years, may be increasing more in the future, and that the world's climate may change as a result. What do you think: Do you think that climate change is happening? If you're not sure, just let me know." Yes, No, Don't know (n = 4025)

Appendix 2: Climate Change in the Chinese Mind 2017

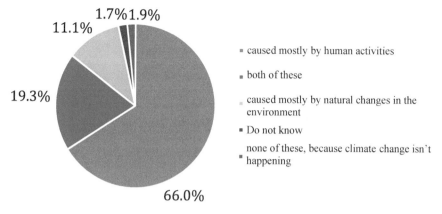

Fig. 22 Assuming climate change is happening, do you think climate change caused mostly by human activities, mostly by natural changes in the environment, or none of these, because climate change isn't happening? (n = 4025)

A5. 66.0% of Respondents Think that Climate Change is Caused "Mostly by Human Activities"

In regard to the cause of climate change, 66.0% of respondents think that climate change is caused "mostly by human activities", while 11.1% say it is due to "natural changes in the environment". 19.3% of respondents think it is caused by both reasons. 1.7% think that climate change "is not happening" (Fig. 22).

A6. 79.8% of Respondents Worry About Climate Change

79.8% of respondents are either "very" (16.3%), or "somewhat" (63.5%) worried about climate change. In contrast, 16.2 and 3.9% of respondents say they are "not very" or "not at all" worried about it, respectively (Fig. 23).

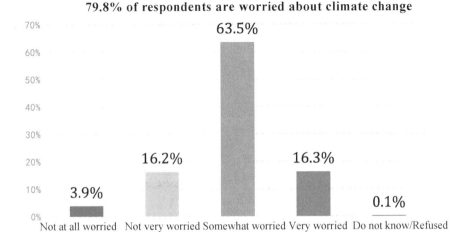

Fig. 23 How worried are you about climate change? Would you say you are very worried, somewhat worried, not very worried, or not worried at all? (n = 4025)

B. Climate Change Impacts

B1. 75.2% of Respondents Have Already Personally Experienced Impacts of the Climate Change

75.2% of respondents say they have already personally experienced impacts of climate change, accounting for over three fourths of all respondents. Only 24.6% hold the opposite view (Fig. 24).

B2. Respondents Think Climate Change Will Harm "Future Generations" and "Plant and Animal Species" the Most

78% of respondents think climate change will have a moderate amount, or a great deal impacts to future generations. 71.7, 51.4 and 31.1% of respondents say such harm will affect plant and animal species, people in China, families and themselves, respectively (Fig. 25).

B3. Most Respondents Think if no Countermeasures are Taken, "Climate Change" and "Disease Epidemics" Will Increase

In next two decades, most respondents think climate change will cause an increase in occurrence of air pollution, followed by disease epidemics, droughts and water,

Appendix 2: Climate Change in the Chinese Mind 2017

75.2% of respondents say they have already personally experienced the impacts of the climate change

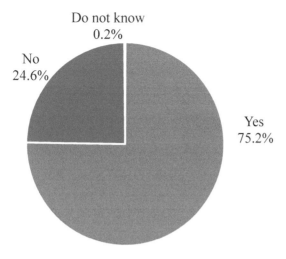

Fig. 24 Have you personally experienced the effects of climate change? (n = 4025)

Respondents think climate change will harm "future generations" and "plant and animal species" the most

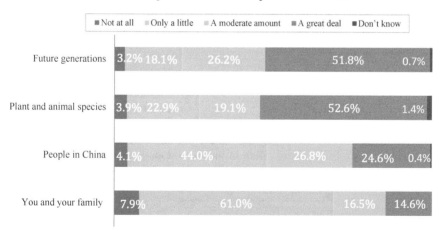

Fig. 25 How much do you think climate change will harm [XXXX]? Would you say a great deal, a moderate amount, only a little, not at all, or do you not know? (n = 4025)

Most respondents think if no countermeasures are taken, "air pollution" and "disease epidemics" will increase

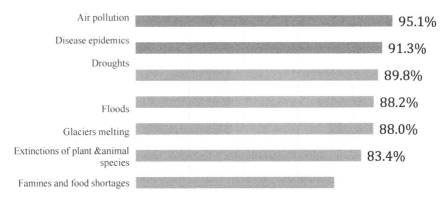

Fig. 26 In China, over the next 20 years, please tell me if you think climate change will cause more or less of the following, if nothing is done to address it? Would you say climate change will cause many more, a few more, a few less, or many less [XXXX], or do you think there will be no difference, or do you not know? (n = 4025). *There are five options for this question—"increase a lot", "increase somewhat", "decrease somewhat", "decrease a lot", "no change has happened". We calculated the percentage of that respondents think a certain phenomenon will increase by adding the percentage of "increase a lot" and "increase somewhat"

shortages, floods, glaciers melting, extinctions of plant and animal species and famines and food shortages, if without any climate change countermeasures in China (Fig. 26).

B4. Most Respondents Worried About Climate Change's Impacts on "Air Pollution" and "Disease Epidemics" the Most

For the question "Which of these impacts are you most worried about", 33.4% of respondents worries about air pollution the most. Others are most worried about disease (29.0%), droughts (10.9%), floods (8.6%) and glaciers melting (6.8%) (Fig. 27).

B5. 72.6% of Respondents Think Climate Change and the Air Pollution are Inter-Related and Have Synergistic Effects on Each Other

In terms of the relationship between climate change and air pollution, 72.6% of respondents think climate change and the air pollution are inter-related, and have synergistic effects on each other. Besides, 14.3% think that climate change leads to air pollution and 12.8% think air pollution leads to the climate change (Fig. 28).

Appendix 2: Climate Change in the Chinese Mind 2017

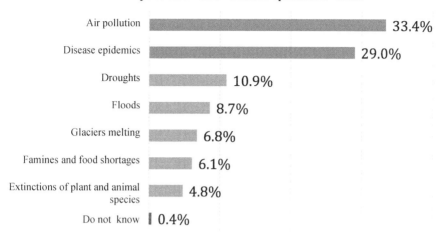

Fig. 27 Which of these impacts are you most worried about? (n = 4025)

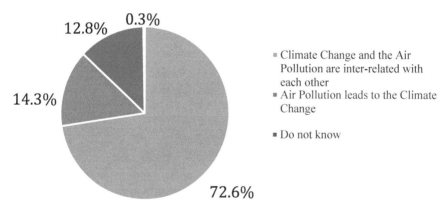

Fig. 28 For the following statements, which one do you agree with? (n = 4025)

Fig. 29 Mitigation and adaptation are two major countermeasures to address climate change. Mitigation refers to efforts to reduce the emission of greenhouse gases, which is the fundamental solution to climate change. Adaptation refers to preparing for climate change impacts like more droughts, floods, and storms. Do you think that mitigation or adaptation is more important, or are they equally important? (n = 4025)

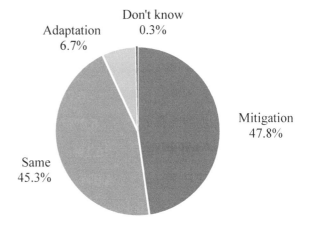

C. Responding to Climate Change

C1. 47.8% of Respondents Think Mitigation and Adaptation are of the Same Importance

47.8% of respondents believe that mitigation is more important than adaptation as countermeasures in address climate crisis. 45.3% of respondents think that mitigation is as important as adaption. Only 6.7% think adaption is more important (Fig. 29).

C2. Respondents Generally Believe That the Government Should do More to Address Climate Change

To address climate change, respondents generally believe that "the government" should do more, followed by "the media" and "environmental NGOs" (Fig. 30).

C3. Respondents Think Air Pollution is the Issue That the Government Should Pay the Most Attention to, Followed by Water Pollution and Ecosystem Protection

When asked about which fields that the central government should pay attention to, among "air pollution, water pollution, climate change, ecosystem protection,

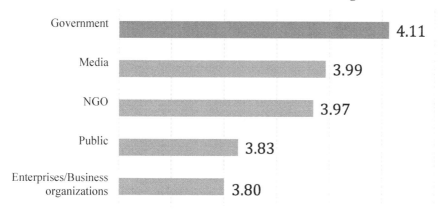

Fig. 30 Do you think each of the following should be doing more or less to address climate change? [Much more, More, Currently doing the right amount, Less, Much less] (n = 4025). *A numerical value from 1 to 5 was assigned to "the least", "less", "just fine", "more", and "the most" in the analysis. The average score each role gets was then calculated

economy development, education, terrorism and health care", over 70% of respondents think all these areas should be given particular attention from the central government.

Averagely, respondents say the issue of air pollution is regarded the most important (3.42), followed by water pollution (3.36), ecosystem protection (3.31), healthcare (3.28) and climate change (3.25) (Fig. 31).

C4. Respondents Think Air Pollution is the Most Critical Problem to be Solved, Followed by Ecosystem Protection and Health Care

Among the aforementioned areas that are of high public attention, 24.3% of respondents think air pollution is the most important, followed by ecosystem protection (18%) and healthcare (17.2%).

8.8% of respondents think climate change is the most important issue, which is even more critical than economic development and anti-terrorism (Fig. 32).

Respondents think air pollution is the issue that the government should pay the most attention to

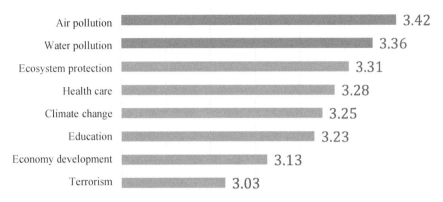

Fig. 31 Here are some issues being discussed by the national government. Do you think each of these issues should be a low, medium, high, or very high priority for President Xi and the national government? (n = 4025). *A numerical value from 1 to 4 was assigned to "low", "moderate", "high", and "very high". An average score each problem got was then calculated

Respondents think air pollution is the most critical problem to be solved

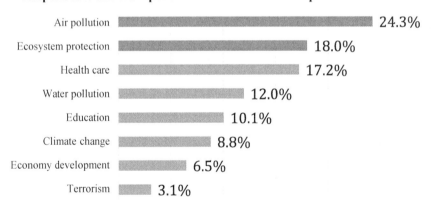

Fig. 32 Of the issues you said should be a priority, which one do you think is most important? (n = 2564)

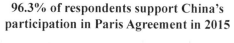
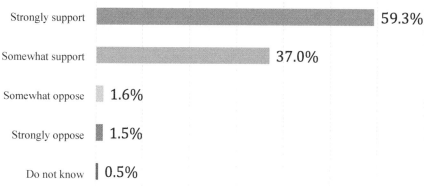

Fig. 33 In 2015, China signed an international agreement in Paris with 195 other countries to limit the pollution that causes climate change. Do you strongly support, somewhat support, somewhat oppose or strongly oppose China's participation in the Paris Agreement to limit the pollution that causes climate change? (n = 4025)

D. Support for Climate Policies

D1. 96.3% of Respondents Support China's Participation in Paris Agreement in the End of 2015

In 2015, China signed the Paris Agreement with 195 other countries. 96.3% of respondents are either "somewhat support" or "strongly support" China's participation in Paris Agreement and among them, 59.3% say they "strongly support" it (Fig. 33).

D2. 94.0% of Respondents Support that China Stays in the Paris Agreement to Honor its Commitments

94.0% of respondents say they support China's decision to stay in the Paris Agreement even if the U.S. withdraw from the Paris Agreement, in which 52.5% say they strongly support (Fig. 34).

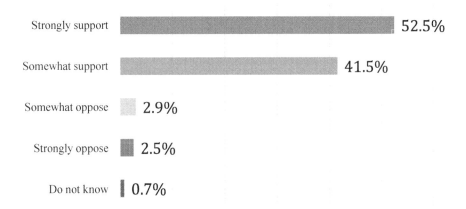

Fig. 34 As the second largest emitter in the world, the U.S. recently announced it will withdraw from the Paris Agreement, but all other countries reaffirmed their pledges. Do you strongly support, somewhat support, somewhat oppose or strongly oppose China's decision to stay in the Paris Agreement to limit the pollution that causes climate change? (n = 4025)

D3. 96.8% of Respondents Support That China Promotes International Cooperation on Climate Change

96.8% of respondents support China's effort to help poorer developing countries mitigate and adapt to climate change, of which over half 54.7% say they "strongly support" it (Fig. 35).

D4. 96.9% of Respondents Support Government's Efforts of the Total Quantity Control on China's Greenhouse Gas Emissions

96.9% of respondents support government's efforts of the total quantity control on China's greenhouse gas emissions and 64.5% are strongly supportive of it (Fig. 36).

D5. Around 90% of Respondents Support the Government's Mitigation Policies

Each policy to mitigate climate change or reduce emissions is "somewhat" supported or "strongly" supported by around 90% of respondents (Fig. 37).

Appendix 2: Climate Change in the Chinese Mind 2017

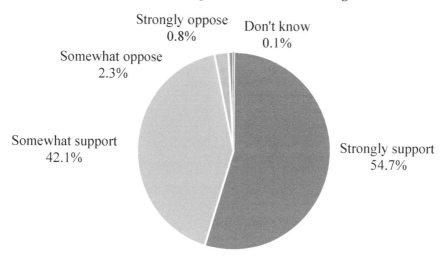

Fig. 35 Do you strongly support, somewhat support, somewhat oppose or strongly oppose China's effort to promote the international cooperation on climate change, i.e. helping poorer developing countries mitigate and adapt to climate change? (n = 4025)

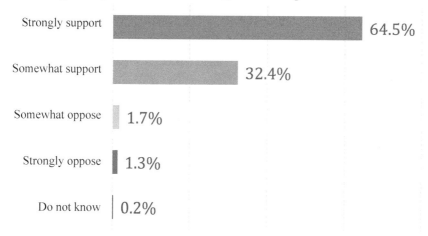

Fig. 36 Do you strongly support, somewhat support, somewhat oppose or strongly oppose the government's efforts of the total quantity control on China's greenhouse gas emissions? (n = 4025)

Around 90% of respondents support the government's mitigation measures

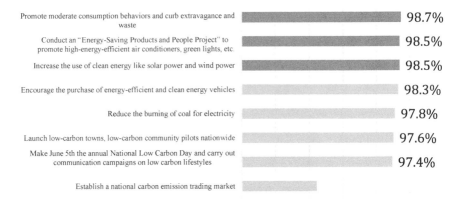

Fig. 37 Do you strongly support, somewhat support, somewhat oppose or strongly oppose each of the following government policies to mitigate or reduce climate change? (n = 4025). *There are four options for this question—"strongly support", "support", "against" and "strongly against". The percentage of "support" and "strongly support" were added up to show how much the respondents support each policy

D6. Over 90% of Respondents Support the Government's Adaptation Policies

Each policy to adapt to climate change is "somewhat" supported or "strongly" supported by over 90% respondents (Fig. 38).

D7. 98.7% of Respondents Support that Schools Should Teach Students About Climate Change

98.9% of respondents support that schools should teach students about the causes, consequences, and potential solutions to climate change and 77.9% "strongly support" it (Fig. 39).

Appendix 2: Climate Change in the Chinese Mind 2017

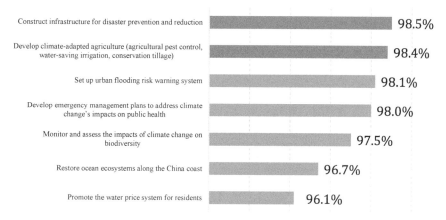

Fig. 38 Do you strongly support, somewhat support, somewhat oppose or strongly oppose each of the following government policies to adapt to climate change impacts? (n = 4025). * There are four options for this question—"strongly support", "support", "against" and "strongly against". The percentage of "support" and "strongly support" were added up to show how much the respondents support each policy

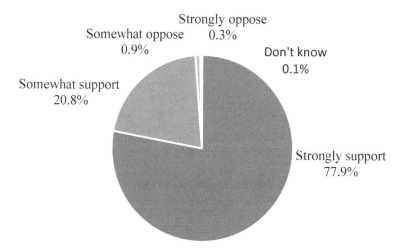

Fig. 39 Do you strongly support, somewhat support, somewhat oppose or strongly oppose that schools should teach our children about the causes, consequences, and potential solutions to climate change? (n = 4025)

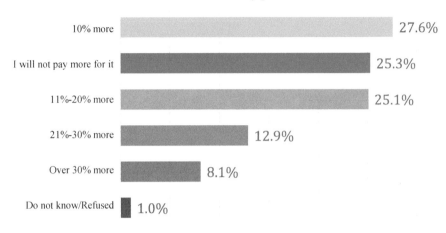

Fig. 40 In general, how much more will you pay for environmental-friendly products which help reduce climate change, like electricity from wind or solar power, energy-efficient appliances, and green housing (i.e. to minimize the resource using and reducing pollution from selecting the construction materials, constructing to decoration and sales), if they cost more? (n = 4025)

E. Enforcement of Climate Change Countermeasures

E1. 73.7% of Respondents are Willing to Pay More for Climate-Friendly Products

When asked if they would be willing to pay more for the climate-friendly products, 73.7% of respondents gave the affirmative answer. Of which 27.6% would pay 10% more at most for such products, representing the largest portion in those who are willing to pay more. 25.1% would pay 11–20% more at most, 12.9% would pay 21–30% more at most, and 8.1% would pay 30% more at most (Fig. 40).

E2. 27.5% of Respondents are Willing to Pay to Offset their Personal Carbon Emissions Completely

When asked if they would like to, 27.5% of respondents are willing to pay to offset their personal emissions completely (RMB 200 yuan per year), accounting for the largest proportion of all respondents. 25% of respondents are willing to pay RMB 100 yuan, 14.7% are willing to pay RMB 50 yuan, and 22.2% are willing to pay RMB 25 yuan (Fig. 41).

Appendix 2: Climate Change in the Chinese Mind 2017

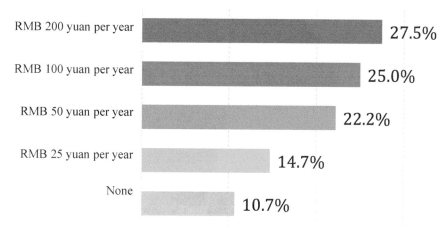

Fig. 41 Everyone generates some pollution when traveling. If offsetting all of your own emissions cost RMB 200 yuan per year, how much would you be willing to pay? (n = 4025)

E3. 46.7% of Respondents Have Used Shared Bikes

When asked if ever used shared bikes, 46.7% of respondents say they have and 53.3% say they have not (Fig. 42).

E4. 92.6% of respondents support using shared bike as a way of travel

When asked if they support using shared bikes, 92.6% say they support it and 6.9% say they do not support. Thus, above 90% of respondents support using shared bike as a way of travel, much more than those who do not (Fig. 43).

E5. 55.6% of Respondents Know the Use of Electricity Generated from Solar PV Installed at Home or in the Company

55.6% of respondents say that they know besides household/company consumption, electricity generated from solar photovoltaic panels can be sold to the State Grid. 44.4% of respondents do not know about it (Fig. 44).

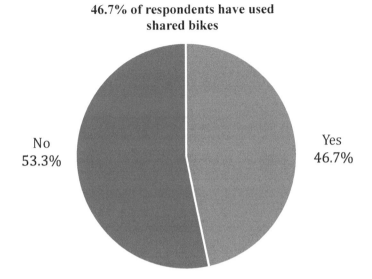

Fig. 42 Have you tried shared bicycle? (n = 4025)

Fig. 43 Do you support the way of shared bicycle? (n = 4025)

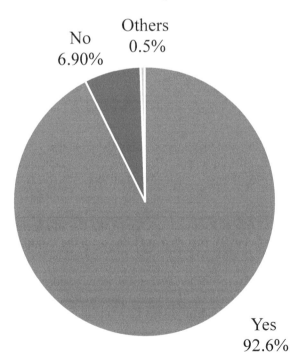

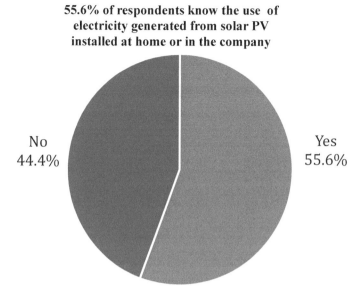

Fig. 44 Have you heard that if you install solar photovoltaic panels you can sell the electricity to the State Grid? (n = 4025)

F. Climate Change Communication

F1. Television and WeChat are Main Channels to Get Climate Change Information

Most respondents are able to access climate change information from various channels and three major information channels are television (83.6%), WeChat (79.4%), and friends and family (68.1%). Newspaper and official pages are also important information channels, as above 50% of respondents choose either of the two. Relatively speaking, respondents are less likely to get climate change information from outdoor billboards, magazines or radio (Fig. 45).

F2. Respondents have Generally Strong Desire to Get Climate Change Related Information

Most respondents have strong desire to learn more about climate change—especially they want to learn more about "climate change impacts" (94.0%) and "climate change solutions" (93.4%). They also care about the "relation between climate change and daily life" and "personal actions to address climate change" (Fig. 46).

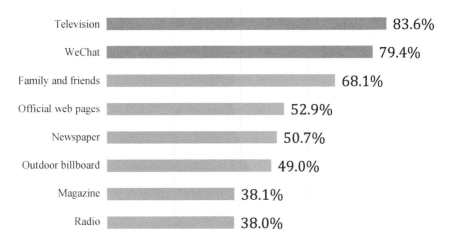

Fig. 45 About how often do you hear or read about climate change from the following: (n = 4025). *The "usage rate" = percentage of "once a year or less" + "many times a year" + "at least once a month" + "at least once a week"

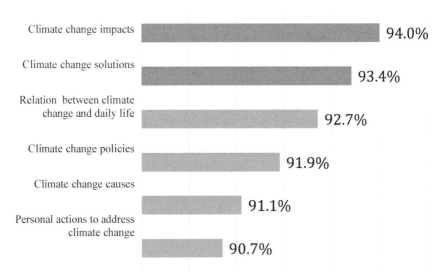

Fig. 46 Do you want to learn more information about each of the following ? (n = 4025). *The percentage of respondents that want to "learn more" = "learn just a little bit more" + "some more" and "much more"

Appendix 2: Climate Change in the Chinese Mind 2017

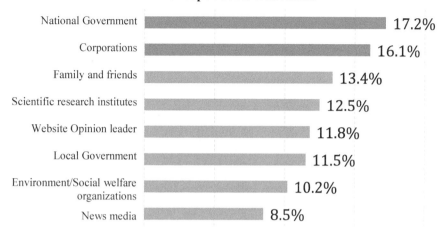

Fig. 47 How much do you trust or distrust the following as a source of information about climate change? (n = 4025). *Numbers used in this figure represent the percentage of respondents choosing "strongly trust"

F3. The Central Government is the Information Source Respondents Trust Most, Followed by Corporations

The most trustful source of information about climate change is the central government 17.2% of respondents selects it as a "strongly trust" source, followed by "corporations (16.1%)", family and friends (13.1%), scientific research institutes (12.5%) (Fig. 47).

F4. Respondents Pay More Attention to Social

Among various types of news, respondents care the social news most (30%), followed by political news (19.9%). About 12.3% of respondents choose environmental news (such as air and water pollution news etc.) as the news they care most (Fig. 48).

F5. 97.7% of Respondents are Willing to Share Climate Change Information

See Fig. 49.

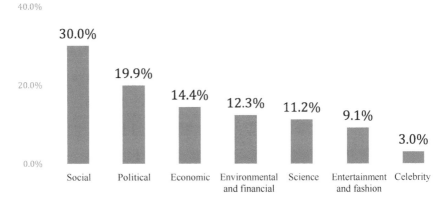

Fig. 48 Which of the following news you care the most about? (n = 4025)

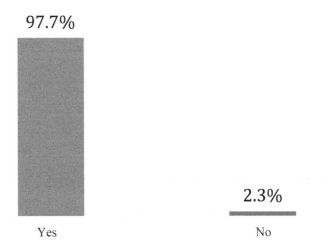

Fig. 49 Are you willing to communicate the climate information with your family and friends? (n = 4025)

Appendix: Sample Demographics

	Frequency	Percentage (%)
Total	4025	100
Male	2054	51.0
Female	1971	49.0
Urban	2337	58.1
Rural	1688	41.9
18–24	515	12.8
25–34	912	22.7
35–44	852	21.2
45–54	906	22.5
55–64	645	16.0
65–70	195	4.8
Primary school or below	358	8.9
Middle school	649	16.1
High school	757	18.8
Technical secondary school	369	9.2
College	842	20.9
Bachelor degree	924	23.0
Master degree or above	126	3.1
Enterprise	1171	29.1
Self-employed	759	18.9
Farming	521	12.9
Public institution	456	11.3
Retired	358	8.9
Unemployed	259	6.4
Students	234	5.8
Administrative organization	99	2.5
Other unemployed	81	2.0
Others	68	1.7
Serviceman	19	0.5

References

In Chinese

Anthony Giddens. *The Politics of Climate Change*. Translated by Rongxiang Cao. Social Sciences Academic Press. 2009.

Yan Bo. "International Negotiations and Domestic Politics. A Double-Level Game Analysis of the U.S. and the Kyoto Protocol". Ph.D. Thesis. Fudan University. 2004.

Brian McNair. *An Introduction to Political Communication*. Translated by Qiyi Yin. Xinhua Publishing House. 2005.

Tuo Cai, Wang Linnan. "Global Governance: A New Mode of Cooperation Adapting to Globalization". Nankai Journal. Issue 2. 2004.

Jinyong Cao, Jue Gong. "Differences, Introspection from Traditional Vision–Climate Environment Changes of Mingyong Village, Deqin County, Yunnan Province, and Villagers' Cognition and Response". Social Sciences in Yunnan. Issue 5. 2014 .

Genying Chang, Fuming Huang, Man Li, Guojing Li. "Global Climate Change Cognition of Villagers from Case Areas of the Loess Plateau and the Southwest of Shandong Province". Geographical Research. Issue 7. 2012.

Rui Chen. "Japanese Media: Perspective of People's Livelihood in Climate Change Reporting". Chinese Journalist. Issue 8. 2007.

Binbin Wang, Qiuyu Gu. "Relationship Study between Chinese Public Climate Perception and Consuming Willingness". China Population, Resources and Environment, 2019,29(9):41-50.

Tao Chen, Hongzuo Xie. "Analysis of Factors Affecting College Students' Willingness to Deal with Climate Change–a Survey Based on 6,643 Questionnaires". Forum of Science and Technology in China. Issue 1. 2012.

Ying Chen. "View Environment and Climate Change Issues from the Security Perspective". World Economics and Politics. Issue 4. 2008.

Weijun Cui, Yu Luo. "The Impact of Urban Residents' Climate Change Risk Cognition on the Choice of Travel Mode–Survey and Analysis Based on 620 Urban Residents." Ecological Economy. Issue 11. 2014.

Bing Dai. "Development of NGOs from the Early 19th Century to 1918". Journal of University of International Relations. Issue 6. 2007.

Thomas Danaldson, Thomas W .Dunfee. *Binding Relationship-A Study of Social Contract Theory of Business Ethics*. Translated by Yuesi Zhao. Shanghai Academy of Social Sciences Press. 2001.

Yongfeng Feng. "Three Key Points of Climate Change Reporting". Chinese Journalist. Issue 8. 2007.

Hanwen Ge. "International Mechanism and Sovereign States in Global Climate Governance". Forum of World Economics and Politics. Issue 3. 2005.
Quansheng Ge, Shaowu Wang, Xiuqi Fang. "Certain Uncertainties in Climate Change Research". Geographical Research. Issue 2. 2012.
Xiaoping Guo. "China's Environmental Image in the Eyes of Western Media–Taking the *New York Times*' "Climate Change" Risk Report. 2000~2009) as an Example". Journalism & Communication. Issue 4. 2010.
Guoping He. "Change and Construction of China's Idea on External Reports: Based on the International Communication Capacity". Shandong Social Sciences. Issue 8. 2009.
Dongmin Hou. "The Status Quo and Prospects of Ecomigration in China's Ecologically Vulnerable Areas". World Environment. Issue 4. 2010.
Baijing Hu. "Innovation and Paradigm Shift of Health Communication Concepts –Dilemma and Solution of Public Communication in the New Media Era". Journal of International Communication. Issue 6. 2012.
Hepeng Jia. "Global Warming. Science Communication and Public Participation—Communication of Climate Change Technology in China". Science Popularization. Issue 6. 2007.
Hepeng Jia. "Several Paths of Climate Change Reporting". Practical Journalism. Issue 10. 2011.
Huanjie Jia. "Who Hijacks the Copenhagen Climate Change Conference?–Critical Discourse of News Quotes. Journalism Lover. Issue 7. 2011.
Shenghua Jia, Honghui Chen. "Review of Stakeholder Definition Methods". Foreign Economies and Management. Issue 5. 2002.
Shiliang Jiang. "Rethinking on Climate Reporting Caused by Snow Disasters". Journalist Monthly. Issue 4. 2008.
Xiaoli Jiang, Li Lei. "Discourse Study of Sino-U.S. Environmental News Reporting—Reports of Four Newspapers of China and the U.S. on the Copenhagen Climate Change Conference". Journal of Southwest Minzu University. Issue 4. 2010 .
Clyde Prestowitz. *Rogue Nation*. Mainly translated by Zhenxi Wang. Xinhua Publishing House. 2004.
Biting Li. "Definition of Key Stakeholders of Coal Enterprises". Theory and Practice of Contemporary Education. Issue 4. 2014.
Huiming Li. "Policy Stances of the EU in International Climate Negotiations". World Economics and Politics. Issue 2. 2010.
Jianmin Li, Yanbing Sun. "International System, National Autonomy and Low-Carbon Economy–China's Policy Choice". Dongjiang Journal. Issue 2. 2011.
Yujie Li. "Cognitive Differences of Urban and Rural Residents in Climate Change, and Construction of Communication Strategy. An Empirical Study Based on 4,169 Questionnaires". Dongyue Tribune. Issue 10. 2013.
Yujie Li. "Information Sources, Channels and Contents – Research on China's Public Climate Communication Strategy Based on the Survey". Journal of International Communication. Issue 8. 2013.
Wei Li. "Evolutionary Logic of International Political Economy". World Economics and Politics. Issue 10. 2009.
Minwang Lin, Liqun Zhu. "Domesticization of International Norms: Impact and Communication Mechanism of Domestic Structure". Journal of Contemporary Asia-Pacific Studies. Issue 1. 2011.
Bing Liu, Qiang Hou. "Research on Domestic Science Communication: Theory and Problem". Studies in Dialectics of Nature. Issue 5. 2004.
Jun Liu. "How to Grasp and Expand the Climate Change Reporting". Chinese Journalist. Issue 8. 2007.
Yuning Liu, Baogui Du. "Main Role of NGOs in Global Climate Negotiations". Journal of Shenyang Agricultural University. Issue 2. 2012.
Hui Luo. "Impact of International NGOs on Global Climate Change Governance – An Analysis Based on the Path of Cognitive Community". Journal of International Relations. Issue 2. 2013.

References

Jing Luo, Jiahua Pan, Enping Li. "College Students' Ethical Orientation of Addressing Climate Change". Impact of Science on Society. Issue 3. 2009.

Margaret E. Keck, Kathryn Sikkink. *Activists Beyond Borders: Advocacy Networks in International Politics*. Translated by Zhaoying Han, Yingli Sun. Peking University Press. 2009.

Jianying Ma. "Internalization of the International Climate System in China". World Economics and Politics. Issue 6. 2011.

Lei Ma. "China's Discourse Strategy of Climate Diplomacy". Journal of Jiamusi Education Institute. Issue 2. 2012.

Yarong Lu, Shufen Chen. "Farmers' Cognition of and Adaptation to Climate Change". Chinese Rural Economy. Issue 7. 2010.

Yong Luo, Yun Gao. "The Conflict of Interests in the Science Communication of Climate Change". Science Popularization. Issue 7. 2012.

Jiahua Pan. "Scientific Controversy and International Political Compromise about National Interests". World Economics and Politics. Issue 2. 2002.

Jiahua Pan, Mou Wang. "New Pattern of International Climate Negotiations and China's Positioning". China Population Resources and Environment. Issue 4. 2014.

Jin Qian, Zhe Qin. "China Issues of Climate Change from the Perspective of Media of Five Countries". Journalism Lover. Issue 8. 2010.

Ru Qu, Ting Lu. "Comparison of *People's Daily* and *Los Angeles Times* on the Reporting of Copenhagen Climate Change Conference". Modern Communication. Issue 10. 2010..

Haijun Ren. "Maintaining National Interests in Climate Change Reporting" Chinese Journalist. Issue 4. 2008.

People's Network. "China's Low-Carbon Energy Strategy and Climate Change Communication". http://ft.people.com.cn/fangtanDetail.do?pid=2731. Last visit date: August 12, 2014.

Xiaofeng Song. "NGOs and Global Climate Governance: Functions and Limitations". Social Sciences in Yunnan. Issue 5. 2012.

Li Sun. "Generation Mechanism and Process Analysis of Public Interest Sectors in China". Comparative Economic & Social Systems. Issue 4. 2006.

Jue Sun and Jiahua Pan. "Doha Round Negotiations and China's Low Carbon Development Road". Environmental Protection. Issue 23.

Dennis T. Avery, S. Fred Singer. "Unstoppable Global Warming: Every 1500 Years". Translated by Wenpeng Lin and Chenli Wang. Scientific and Technical Documentation Press. 2008.

Zhixin Tan. "Shandong Farmers' Cognition and Adaptation to Climate Change". Chongqing Social Sciences. Issue 3. 2011.

Binbin Wang. "Public Participation in Addressing Climate Change". World Environment. Issue 1. 2014.

Chunyang Wang, Fucai Huang. "Empirical Research on the Definition and Classification of Stakeholders of Village Heritage Sites–Taking Watchtowers and Villages in Kaiping as an Example". Tourism Tribune. Issue 8. 2012.

Jiafu Wang. *International Strategy*. Higher Education Press. 2005.

Mei Jin. "Interaction between Non-State Actors and Sovereign States in International Climate Governance". Shangqing Journal. Issue 33. 2011.

Jinna Wang, Yongjie Wang, Ying Zhang, Xiao Zhang, Caixia Yang, Baofa Jiang. "A Survey on College Students' Cognitive Status of Climate Change". Journal of Environment and Health. Issue 7. 2012.

Wenjun Wang. "Low-Carbon Economy: Lessons from Foreign Experience and China's Development". Journal of Northwest A&F University. Issue 6. 2009.

Zhan Wang, Hailiang Li. "A Review of Western Climate Change Communication Research". Southeast Communication. Issue 3. 2011.

Zizhong Wang. *Climate Change: Political Kidnapping of Science*. China Financial & Economic Publishing House. 2010.

Werner J. Severin, James W. Tankard. *Communication Theories: Origins, Methods and Uses in the Mass Media*. Translated by Zhenzhi Guo, Ying Meng et al. Huaxia Publishing House. 2001.

Ulrich Beck. *Risk Society*. Translated by He Bowen. Yilin Press. 2004.
Xinhuanet. "Records on Premier Wen Jiabao's Attending the Copenhagen Climate Change Conference" http://news.xinhuanet.com/world/2009-12/24/content_12700839.htm. 2009.
Xinmin.cn. "Environmental Protection Organization: Adapting to Climate Change is More Realistic and More Urgent Than Mitigation". http://finance.ifeng.com/a/20140506/12272246_0.shtml. Last Visit Date. May 6, 2014.
Lanlan Xiao. "Farmers' Cognition of Climate Change–Empirical Research and Analysis of Farmers in Some Areas of Shandong Province". Journal of Qingdao Agricultural University (Social Science Edition. Issue 2. 2013.
Yungeng Xie. "Approaches, Issues and Perspectives of Risk Communication Research". New Media and Society (third series). Issue 11. 2012.
Nihong Xie. "Review and Reflection of Stakeholder Research". Friends of Accounting. Issue 10. 2009.
Guangqing Xu. "Statistical Analysis of Climate Change Awareness of Enterprise Managers". China Population Resources and Environment. Issue 7. 2011.
Qi Xu. "Increase the 'Thickness' of Climate Change Reporting". Chinese Journalist. Issue 8. 2007.
Yi Yang. "Domestic Constraints, International Image and China's Climate Diplomacy". Social Sciences in Yunnan. Issue 1. 2012.
Lan Yu, Rui Zhou. "Focus on Durban: BASIC's Response to" Split "Rumor at a 'Super Specification' Press Conference". China News Network (http://www.chinanews.com/gj/2011/12-07/3513080.shtml. 2011.
Lihua Yang, Ji Ma, Shijing Yan. "Reflection on China's Media Climate Reporting". Journal of Southwest Minzu University. Issue 8. 2010.
Qingtai Yu. "The Keynote Speech at the Economic Monthly Talk Hosted by the China Center for International Economic Exchanges". http://www.china.com.cn/zhibo/2010-02/24/content_1945 3171.htm. 2010.
Yaru Yun, Xiuqi Fang, Qing Tian. "Preliminary Analysis of Climate Change Perception of Rural Residents—Taking Mohe County in Heilongjiang Province as an Example". Advances in Climate Change Research. Issue 5. 2009.
Jiequan Zhai, Zhijian Yang. "Analysis of Scientific Communication Concept". Journal of Beijing Institute of Technology. Issue 8. 2008.
Haibin Zhang. "China's Position in International Climate Change Negotiations: Continuity, Change and Causes". World Economics and Politics. Issue 10. 2006.
Haibin Zhang. "China and International Climate Change Negotiations". International Politics Quarterly. Issue 1. 2007.
Haibin Zhang. "Special Study: Climate Change and China's National Security". International Politics Quarterly. Issue 4. 2015.
Lei Zhang. "China's Dilemma in International Climate Politics: A Micro-Level Combing". Teaching and Research. Issue 2. 2010.
Xiaoyu Zhang, Wenjing Shi. *A Review of Global Climate Change Disputes*. China Center for Energy and Development (CCED) at National School of Development (NSD), Peking University. 2010.
Zhang Zhang. "Trilogy of Western TV Media Reporting about Major Events Involving China–Taking the Copenhagen Climate Change Conference as an example". Southeast Communication. Issue 4. 2010.
Baowei Zheng, Binbin Wang, Yujie Li. *Win-Win Cooperation in Climate Communication–Role and Influence of Chinese Government, Media and NGOs in the Post-Copenhagen Era. Collection of Papers on Journalism and Communication* (Issue 24). *Economic Daily Press*. 2010.
Baowei Zheng, Binbin Wang, Yujie Li. *The Theory and Practice of Climate Change Communication*. People's Daily Press. 2011.
Baowei Zheng, Binbin Wang. *Climate Communication Strategy and Skills of Chinese Government, Media and NGOs. Collection of Papers on Journalism and Communication* (Issue 27). *Economic Daily Press*. 2011.

References

Baowei Zheng, Binbin Wang. "Media Communication Strategy of Four Types of Low Carbon People in Chinese Cities". Journal of International Communication. Issue 8. 2013.

Baowei Zheng, Binbin Wang. "Context, Opportunities and Challenges of China's Climate Communication Research". Dongyue Tribune. Issue 10. 2013.

China News Network. "Xie Zhenhua's Frank Dialogue with International NGOs on Climate Change". http://www.chinanews.com/gn/2010/10-08/2573501.shtml. Last visit time: October 8, 2010.

China News Network. "Developed Countries Have Repeatedly Won the Fossil of the Day for Hindering Climate Negotiations, and China's Stance Draws Attention". http://finance.sina.com.cn/chanjing/cyxw/20111201/230910919458.shtml. Last visit time: December 1, 2011.

Guiyang Zhuang. "International Climate Governance in the Post-Kyoto Era and China's Strategic Choice". World Economics and Politics. Issue 8. 2008.

Liang Dong. "How IPCC Affects International Climate Negotiations–An Analysis Based on the Cognitive Community Theory". World Economics and Politics. Issue 8. 2014.

Yangmei Han, Weidong Zhuge. "International Frontier Status Quo and Trend of Climate Communication Research–A Case Study of *The Public Understanding of Science* and *Science Communication*. 2006-2015. Science Popularization. Issue 4. 2017.

Keqiang Li. Government Work Report 2017. Chinese Government Website. http://www.gov.cn/guowuyuan/2017-03/16/content_5177940.htm. Last visit time: March 16, 2017 .

Mancur Olson. *The Logic of Collective Action*. Translated by Chen Yu, Guo Yufeng, and Li Chongxin. Shanghai People's Publishing House. 1995.

Qian Qiu. "Process of the Kyoto Protocol's Entry into Force". Journal of Party and Government Cadres. Issue 11, 2005.

Huiming Li. "Order Transformation—Root Causes for Hegemony Decline, Global Climate Politics Governance System Fragmentation and Leadership Lack". Journal of PLA Nanjing Institute of Politics. Issue 6. 2014.

Yan Bo, Zhimin Chen. "Weakening of EU Leadership in Global Climate Change Governance". International Studies. Issue 1. 2011.

Liang Dong. "Conference Diplomacy, Negotiation Management and Paris Climate Conference". Foreign Affairs Review. Issue 2. 2017.

Chinese Government Website. Outline of the 12th Five-Year Plan for National Economic and Social Development. http://www.gov.cn/2011lh/content_1825838.html. Last visit time: March 16, 2011.

Huiming Li. "The Paris Agreement and the Transformation of Global Climate Governance System". World Outlook. Issue 2. 2016.

Guiyang Zhuang, Weiduo Zhou. "Transformation of Global Climate Governance Mode and China's Contributions". The Contemporary World. Issue 1. 2016.

Ji, Chen Junfeng Li. "A Long Way to Go in Implementation of the Paris Agreement". Environment Economy. Issue 9, 2016.

Yongfu Huang. "The Latest Achievements in Global Climate Change Governance". People's Tribune. Issue 4. 2016.

Maorong Zhou. "Opportunities, Challenges and Countermeasures for China to Implement the Paris Agreement". Journal of Environmental Economics. Issue 2. 2016.

Binbin Wang. "China's Climate Communication Strategy Based on Double-layer Game Framework". Ph.D. Thesis. Renmin University of China. 2015.

Xiang Gao. "China's Progress and Prospects for South-South Cooperation on Climate Change Response". Journal of Shanghai Jiaotong University. Issue 1. 2016.

Liang Dong. "The Impact of the 2030 Agenda for Sustainable Development on Environmental Governance in China and the World". China Population, Resources and Environment. Issue 1. 2016.

Qimin Chai, Xiang Gao, Huaqing Xu. *BASIC: Climate Governance from Copenhagen to Paris*. China Planning Press. 2016.

Helen Milner. *Interests, Institutions and Information: Domestic Politics and International Relations*. Translated by Qu Bo. Shanghai People's Publishing House. 2015.

Jing Lu. "Current Systemic Dilemma and Reform of Global Governance". Foreign Affairs Review. Issue 1. 2014.
Hong Ding. Globalization, Global Governance and International NGOs. Forum of World Economics & Politics. Issue 6. 2006.
Qingcai Liu, Nongshou Zhang. "Role of NGOs in Global Governance". International Studies. Issue 1. 2006.
Jin Xu, Chang Liu. "Chinese Scholars' Study of Global Governance". Quarterly Journal of International Politics. Issue 1. 2013.
Lan Xue, Hanzhi Yu. "Global Governance with Public Management Paradigm Nature—Analysis Based on the 'Problem-Subject-Mechanism' Framework". Social Sciences in China. Issue 11. 2015.
Yaqing Qin. "Global Governance Failure and Order Concept Reconstruction". World Economics and Politics. Issue 4. 2013.
Li Li. *Public-Private Partnerships in Global Climate Governance.* Current Affairs Press. 2013.
Yan Bo, Xiang Gao. "Changes in Climate Governance Mechanisms of China and the World". Shanghai People's Publishing House. 2017.
Longbiao Zhong, Jun Wang. "Two-level Game in China's Peaceful Rise –Review on *China: Fragile Superpower: How China's Internal Politics Could Derail Its Peaceful Rise*". American Studies Quarterly. Issue 4. 2007.
Shenyu Wang. "From Impact and Participation to Common Governance–Historical Leap and Enlightenment of Stakeholder Theory Development". Journal of Xiangtan University. Issue 6. 2008.
Gang Zhang. "Chinese Scholars' Two-Level Game Research". Journal of Liaoning Provincial College of Communications. Issue 4. 2009.
Hao Meng. "China's Low Carbon Development Strategy for Climate Change". 15th Annual Meeting of Chinese Association of Productivity Science and Seminar for WAPS Academicians. Beijing. 2009.
Xinhua News Agency. "View New Highlights of China's Ecological Civilization from the Bonn Climate Change Conference". http://www.xinhuanet.com/world/2017-11/16/c_1121965658.html.
Yingchun Xu. "Green Myth: The Concept, Field, Method and Framework of Environmental Communication Research". Overseas Report of China Media. Issue 1. 2013.
Wei Su. "Cognitive Popularization of Climate Change at the Copenhagen Climate Change Conference". 2009. China Network. http://www.china.com.cn/news/2009-12/30/content_19157144.html. Last visit time: December 30, 2009.
Zheng Yu. "Public Opinion War Behind International Issues of Climate Change". Chinese Journalist. Issue 2. 2010.
Tao Liu. "Rhetoric Theory of New Social Movements and Climate Communication". Journal of International Communication. Issue 8. 2013.
Oxfam, Greenpeace, Chinese Academy of Agricultural Sciences. Climate Change and Poverty-China Case Study Report. http://www.oxfam.org.cn/uploads/soft/20130428/1367143945.pdf. 2009 .
Honghui Chen, Shenghua Jia. "Empirical Analysis of Three-Dimensional Classification of Corporate Stakeholders". Economic Research Journal. Issue 24. 2004.
Xinhua Daily Telegraph. "Work Hard to Build a New Great-Power Relationship between China and the U.S.". http://news.xinhuanet.com/mrdx/2014-07/10/c_133473075.html. Last visit time: July 9, 2014.
Jie Feng. "Global Collaboration of NGOs: The Largest Coal Enterprise's High-Profile Response to 'If This is a War, Any Communication Will Not Happen'". Southern Weekend. http://www.infzm.com/content/100313. Last visit time: May 1. 2014.
Shuyong Guo. International Image in the Growth of Great Power. International Forum. Issue 6. 2005.

Changhe Su. Transnational Relations and Domestic Politics–A Study of International Relations from the Perspective of Comparative Politics and International Political Economy. American Studies Quarterly. Issue 4. 2003.
Hua Wang. "Partnership in Governance: Cooperation between Government and NGOs". Social Sciences in Yunnan. Issue 3. 2003.
Lina Wang, Yue Zhou, Ling Huang. "Review: Joys and Worries from Marrakesh Climate Change Conference". Xinhua News Agency: http://news.xinhuanet.com/2016-11/17/c_1119935504.html. Last visit time: November 17, 2016.
Xiaoyu Li, Lan Yu. "New President Trump Stirs Marrakesh Climate Change Conference". China News: http://www.chinanews.com/gj/2016/11-09/8058275.shtml. Last visit time: November 9, 2016.
Suyan Lu. "The United Nations Issues a Communiqué to Celebrate the Entry into Force of the Paris Agreement". Xinhua News Agency: http://news.xinhuanet.com/world/2016-11/04/c_1119850305.html. Last visit time: November 4, 2016.
Xiaochun Lin. "News Analysis: Trump Administration's Energy Plan Is Disconcerting". Xinhua News Agency: http://news.xinhuanet.com/2017-01/22/c_1120362431.html. Last visit time: January 22, 2017.
Miao Zhang. "The United Nations: Gap Between INDC and Emission Targets". Xinhuanet: http://news.xinhuanet.com/world/2015-11/06/c_1117067824.html. Last visit Time: November 6, 2015.
"Keynote Speech Made by President Xi Jinping at the Opening Ceremony of the Annual Meeting of the World Economic Forum 2017". Xinhua News Agency: http://news.xinhuanet.com/fortune/2017-01/18/c_1120331545.html. Last visit Time: January 18, 2017.
China News. "International Media's Positive Evaluation of Xi Jinping's Speech in Davos: Showing the Image of Great Power". http://www.chinanews.com/gn/2017/01-19/8128979.shtml. Last visit time: January 19, 2017.
Zexi Hu, Zhang Penghui, Li Yingqi. "Xi Jinping's Speech in Davos Arouses Hot Discussion and International Community Praises China's Responsibility". China Network: http://www.china.com.cn/news/world/2017-01/18/content_40130863.html. Last visit time: January 18, 2017.
Jianmin Li. "Xi Jinping: Working Together to Create a Green, Healthy, Intellectual and Peaceful Silk Road". Xinhua News Agency: http://news.xinhuanet.com/politics/2016-06/22/c_1119094645.html. Last visit time: June 22, 2016.
Junfeng Li, Ji Chen, Xiu Yang. "Win-win Cooperation–Comments on China's INDC". National Center for Climate Change Strategy and International Cooperation. http://www.ncsc.org.cn/article/yxcg/zlyj/201507/20150700001487.shtml. Last accessed. July 1, 2015.
Haibin Zhang. "A Long Way to Go in Implementation of the Paris Agreement". International Herald Leader. 2016.
Shumin An, ShiqiuZhang. "China's Climate Governance Challenges and Responses under the Paris Agreement". Environmental Protection. Issue 22. 2016.
Bin Wu. "High-end Interview I–China's Special Representative Xie Zhenhua Responsible for Climate Change Matters: China Will Present Solutions for All Climate Negotiation Issues". Nanfang Metropolis Daily. March 14, 2017.
Linfei Zhi. "Xinhua Review: China and the U.S. Should Develop Relations Seizing the Opportunity". Xinhua News Agency: http://news.xinhuanet.com/world/2017-03/16/c_129511380.html. Last visit time: March 15, 2017.
Xiaomin Wang. "NGOs and Sustainable Development". Theory Monthly. Issue 10. 2008.
Fang Wang. "Vice President of the World Bank: China's 'INDC' Is a World Example". China Network: http://news.china.com.cn/world/2015-12/10/content_37282503.html. Last visit time: December 10, 2015.
China News. "Xie Zhenhua's Comment on Bonn Climate Change Conference: Reflecting Win-win Cooperation, and Laying a Good Foundation". http://www.chinanews.com/cj/2017/11-18/8380001.shtml.
Official website of Greenpeace. http://www.greenpeace.org.cn/site/climate-energy/2016/powerlab/incubator.php.

Website of Global Green Leadership. http://www.chinagoinggreen.org.
Sohu News. "Chinese Enterprise Meeting of United Nations Climate Conference in Bonn". http://www.sohu.com/a/203605923_480207. Last visit time: November 10, 2017.
People's Daily Online. "President Fatih Birol of International Energy Agency Delivers a Speech at the Closing Bonn Climate Change Conference on November 16, 2017, Frequently Praising China". http://world.people.com.cn/nl/2017/1117/cl002-29652779.html. Last visit time: November 17, 2017.
Chengchuan Tian et al. *A World of Taiji:Comparative Study of Climate Change Strategy Between China and the U.S.* People's Publishing House. 2017.
Songli Zhu, Xiang Gao. *From Copenhagen to Paris: Changes and Development of International Climate System.* Tsinghua University Press. 2017.

In English

Adam, S., David, N.H. 2012. "Framing Climate Change." *Journalism Studies* 3 (2).
Akerlof, K., R. Debono, P. Berry, A. Leiserowitz, C. Roser-Renouf, K. Clarke, A. Rogaeva, M. Nisbet, M. Weathers, and E. Maibach. 2010. "Public Perceptions of Climate Change as a Human Health Risk: Surveys of the U.S., Canada and Malta." *International Journal of Environmental Research and Public Health* (7).
Andorno, R. 2004. "The Precautionary Principle. A New Legal Standard for a Technological Age." *Journal of International Biotechnology Law* (1).
Balmford A., A. Manica, L. Airey, L. Birkin, A. Oliver, and J. Schleicher. 2004. "Hollywood, Climate Change, and the Public." *Science305* (1713).
Becker, M. 2005. "Accepting Global Warming as a Fact." *Nieman Reports,* 59 (4).
Bernard D. Goldstein. 1999. "The Precautionary Principle and Scientific Research are not Antithetical." *Environmental Health Perspectives* 107 (12).
Boykoff, M., and Boykoff, J. 2004. "Balance as Bias. Global Warming and the US Prestige Press." *Global Environmental Change* 14 (2).
Boykoff, M. 2005. "The Disconnect of News Reporting from Scientific Evidence." *Nieman Reports* 59 (4).
Bulkeley, H. 2000. "Common Knowledge? Public Understanding of Climate Change in Newcastle, Australia." *Public Understanding of Science* (9).
Campbell, P. 2011. "Understanding the Receivers and the Reception of Science's Uncertain Messages." *Philosophical Transactions of the Royal Society* (369).
Carpenter, S. C, Walker, B. H., Anderies, M., and Abel, N. 2001. "From Metaphor to Measurement: Resilience of What to What?" *Ecosystems* 4.
Carvalho, A. 2005. "Representing the Politics of the Greenhouse Effect." *Critical Discourse Studies2* (1).
Charkham, J. 1992. "Corporate Governance. Lessons from Abroad. *European Business Journal* 4 (2).
Clarkson, M. 1995. "A Stakeholder Framework for Analyzing and Evaluating Corporate Social Performance." *Academy of Management Review* 20 (1).
Cooney, R. 2003. "The Precautionary Principle in Natural Resource Management and Biodiversity Conservation: Situation Analysis, IUCN" https://www.pprinciple.net/publications/sa.pdf.
Corbett, J. B. and Durfee, J. L. 2004. "Testing Public (un) Certainty of Science. Media Representations of Global Warming. *Science Communication* 26 (2).
Covello, V. T., Slovic P., and Von Winterfeldt, D. 1986. "Risk Communication. a review of literature." *Risk Abstracts* 3 (4).
Cox, Robert. 2010. *Environmental Communication and the Public Sphere.* London. Sage Publications.

References

Craig Idso, and S. Fred Singer. 2009. *Climate Change Reconsidered:* 2009 *Report of the Nongovernmental Panel on Climate Change.* Chicago. IL: The Heartland Institute.

David, M., HughLaFollette, eds. 2003. *Whistleblowing: The Oxford Handbook of Practical Ethics.* New York, Oxford. Oxford University Press.

De Sadeleer, N. 2002. *Environmental Principles.* Oxford University Press.

Douglas, M., and Wildavsky, A. 1982. *Risk and Culture: An Essay on the Selection of Technological and Environmental Dangers.* Los Angeles: University of California Press.

David Archer, and Stefan Rahmstorf. 2010. *The Climate Crisis, Climate Change so far.* Cambridge University Press.

David Gee, and Sofia Guedes Vaz. 2001. *Late Lessons from Early Warning: the Precautionary Principle* 1896–2000. Copenhagen. European Environment Agency.

Donner, S. 2011. "Making the Climate a Part of the Human World." *Bulletin of the American Meterological Society.*

Etkin, D., and E. Ho. 2007. "Climate Change: Perceptions and Discourses of Risk." *Journal of Risk Research.*

Featherstone, H., E. Weitkamp, K. Ling, and F. Burnet. 2009. "Defining Issue-Based Publics for Public Engagement. Climate Change as a Case Study." *Public Understanding of Science.*

Fisher, A. C., and U. Narain. 2002. "Global Warming, Endogenous Risk and Irreversibility." *Working Paper, Department of Agricultural and Resource Economics, UC Berkeley.*

Frederick W. C. 1988. "Business and Society, Corporate Strategy, Public Policy, Ethics." *McGraw 2 Hill Book Co.*

Freeman Edward R. 1984. *Strategic management: A stakeholder approach.* Boston: Pitman.

Freimond, C. 2007. "Global warming reaches the boardroom. *Communication World.*

Funtowicz, S. 0., and Ravetz, J. R. 1990. *Uncertainty and Quality in Science for Policy.* Dordrecht: Kluwer Academic Publishers.

Gelbspan, R. 2005. "Disinformation, financial pressures, and misplaced balance." *Nieman Reports* 59 (4).

Herriman, J., A. Atherton, and L. Vecellio. 2011. "The Australian Experience of World Wide Views on Global Warming. The First Global Deliberation Process." *Journal of Public Deliberation* 7 (1).

Hulme, M. 2009. *Why We Disagree About Climate Change: Understanding Controversy, Inaction and Opportunity.* Cambridge: Cambridge University Press.

Kahlor, L., and S. Rosenthal. 2009. "If We seek, Do We Learn? Predicting Knowledge on Global Warming." *Science Communication* (30).

Knight, F. H. (1921) .*Risk, Uncertainty and Profit.* Boston: Houghton Mifflin.

Koteyko, Nelya, Thelwall, Mike, Nerlich, Brigitte. 2010. "From Carbon Markets to Carbon Morality. Creative Compounds as Framing Devices in Online Discourses on Climate Change Mitigation." *Science Communication* 32 (1).

Leiserowitz, A. 2004. "Before and After the Day After Tomorrow. A U. S. Study of Climate Change Risk Perception." *Environment* 46.

L. Mark Berliner. 2003. "Uncertainty and Climate Change. *Statistical Science,* 18 (4).

Lorenzoni, I., and N. Pidgeon. 2006. "Public Views on Climate Change. European and USA Perspectives." *Climatic Change* 77.

Lorenzoni, I., S. Nicholson-Cole., and L. Whitmarsh. 2007. "Barriers Perceived to Engaging with Climate Change among the UK Public and Their Policy Implications." *Global Environmental Change* 17.

Lowe, T., K. Brown., and S. Dessai. 2006. "Does Tomorrow Ever Come? Disaster Narratives and Public Perceptions of Climate Change." *Public Understanding of Science* 15.

Marjolein B. A. VAN Asselt, and Ellen VOS. 2006. "The Precautionary Principle and the Uncertainty Paradox." *Journal of Risk Research 9* (4).

Matthew C. Nisbet. 2009. "Communicating Climate Change. Why Frames Matter for Public Engagement." *Environment: Science and Policy for Sustainable Development* 51 (2).

Mitchell, A., and Wood, D. 1997. "Toward a Theory of Stakeholder Identification and Salience. Defining the Principle of Who and What really Counts." *Academy of Management Review* 22 (4).

Moser, S. C., and Dilling, L. 2004. "Making Climate Hot." *Environment* 46 (10).

Moyers, B. 2005. "How do we cover penguins and politics of denial." *Nieman Reports* 59 (4).

Nisbet, M. 2009. "Communicating Climate Change. Why Frames Matter for Public Engagement." *Environment* (51).

Ockwell, D., L. Whitmarsh, and S. O'Neill. 2009. "Reorienting Climate Change Communication for Effective Mitigation. Forcing People to be Green or Fostering Grass-Roots Engagement?" *Science Communication* (30).

Ohe, M., and S. Ikeda. 2005. "Global Warming. Risk Perception and Risk-Mitigation Behaviour in Japan." *Mitigation and Adaptation Strategies for Global Change* (10).

Olausson, U. 2011. "We're the Ones to Blame: Citizens' Representations of Climate Change and the Role of the Media." *Environmental Communication* (5).

Ralph B. Alexander. 2009. "Global Warming False Alarm, The fuss about C02" *Canterbury Publishing.*

Robert Cox. 2006. "Environmental Communication and the Public Sphere." *SAGE Publications.*

Robert D. Putnam. 1988. "Diplomacy and Domestic Politics. The Logic of Two-level Games." *International Organization* 42 (3).

Russill, C. 2007. "Truth Claims in Climate Change. An Inconvenient Truth as Philosophy of Communication." Paper presented at the meeting of the International Communication Association, San Francisco, CA.

Ryghaug, M., K. Sorensen, and R. Naess .2011. "Making Sense of Global Warming. Norwegians Appropriating Knowledge of Anthropogenic Climate Change." *Public Understanding of Science.* (20).

Patrick. Michaels, and Robert C. Balling JR. 2008. *Climate of Extremes, Pervasive Bias and Climate Extremism.* Cato Institute.

Sampei, Y., and M. Aoyagi-Usui .2009. "Mass-Media Coverage, its Influence on Public Awareness of Climate Change Issues, and Implications for Japan's National Campaign to Reduce Greenhouse Gas Emissions." *Global environmental change* (19).

Schweitzer, S., J. Thompson, T. Teel, and B. Bruyere. 2009. "Strategies for Communication about Climate Change Impacts on Public Lands." *Science Communication* (31).

Sundblad, E., A. Biel, and T. Garling. 2008. "Knowledge and Confidence in Knowledge about Climate Change among Experts, Journalists, Politicians, and Laypersons." *Environment and Behaviour* (41).

Sussane C. Moser. 2010. "Communicating Climate Change. History, Challenges, Process and Future Directions." *Wiley Interdisciplinary Reviews: Climate Change* (1).

Tolan, S. and Berzon, A. 2005. "Global Warming. What's Known vs. What's Told." *Nieman Reports* 59 (4).

Uggla, Y. 2008. "Strategies to Create Risk Awareness and Legitimacy. The Swedish Climate Campaign." *Journal of Risk Research (11).*

Van Asselt & Rotmans, J. 2002. "Uncertainty in Integrated Assessment Modelling. from Positivism to Pluralism." *Climate Change.*

Von Storch, H., and Krauss, W. 2005. "Culture contributes to perceptions of climate change." *Nieman Reports* 59 (4).

W. E. Walker et al. 2003. "Defining Uncertainty A Conceptual Basis for Uncertainty Management in Model-Based Decision Support. *Integrated Assessment* 4 (1).

Wheeler D., and Maria S. 1998. "Including the Stakeholders. the Business Case." *Long Range Planning* 31 (2).

Wolf, J., and S. Moser. 2011. "Individual Understandings, Perceptions, and Engagement with Climate Change. Insights from In-Depth Studies across the World." *Wiley Interdisciplinary Reviews*: *Climate Change* (2).

References

Zhao, X., A. Leiserowitz, E. Maibach, and C. Roser-Renouf. 2011. "Attention to Science/ Environment News Positively Predicts and Attention to Political News Negatively Predicts Global Warming Risk Perceptions and Policy Support." *Journal of Communication* (61).

Wang, B. B., Shen, Y. T, Jin, Y. Y. 2017. "Measurement of Public Awareness of Climate Change in China: Based on a National Survey with 4, 025 Samples." *Chinese Journal of Population Resources and Environment* 15 (4).

Zartman, W. 1994. *International Multilateral Negotiation: Approaches to the Management of Complexity.* San Francisco: Jossey-Bass Publishers.

Young, O. 1991. "Political Leadership and Regime Formation: On the Development of Institutions in International Society." *International Organization,* (45).

UNFCCC. 2005. "Ad Hoc Working Group on Further Commitments for Annex I Parties under the Kyoto Protocol." http://unfccc.int/bodies/body/6409.phf.

Arker, C. F., and C. Karlsso, and M. Hjerpe, et al. 2012. "Fragmented Climate Change Leadership. Making Sense of the Ambiguous Outcome of COP15." *Environmental Politics* (3).

Hilton, I., and O. Kerr. 2016. "The Paris Agreement: China's New Normal Role in International Climate Negotiations." *Climate Policy* (1).

Clemencon, R. 2016. "The Two Sides of The Paris Climate Agreement: Dismal Failure or Historic Breakthrough." *Journal of Environment & Development* (1).

Sabel, C. F., D. G. Victor. 2015. "Governing Global Problem Under Uncertainty: Making Bottom-up Climate Policy Work." *Climate Change* (10).

Andresen, S. 2007. "Key Actors in UN Environmental Governance: Influence, Reform and Leadership." *International Environment Agreements Politics Law Econ* (7).

Dong, L. 2017. "Bound to lead? Rethinking China's Role after Paris in UNFCCC Negotiations." *Chinese Journal of Population Resources and Environment* (2).

Ahlquist, J. S., M. Levi. 2011. "Leadership. What it Means, What it does, and What We Want to Know about It." *Annual Review of Political Science* (14).

Parker, C. F., C. Karlsson, and M. Hjerpe. 2015. "Climate Change Leaders and Followers." *International Relations* (4).

G. John I ken berry. 2011. "The Future of the Liberal World Order: Internationalism After America." *Foreign Affairs* 90 (3).

World Bank. 1996. World Bank Participation Sourcebook. Washington D. C. : World Bank.

IPCC, 2014 . "Approved Summary for Policymakers, IPCC Fifth Assessment Synthesis Report. http://www.ipcc.ch/pdf/assessment-report/ar5/syr/SYR_AR5_SPMcorrl.pdf.

IPCC, 2014. "Impacts, Adaptation and Vulnerability, The Working Group II Report." http://www.ipcc.ch/pdf/assessment-report/ar5/wg2/ar5_wgII_spm_zh.pdf.

Center for Research on Environmental Decisions. 2009. The Psychology of Climate Change Communication: A Guide for Scientists, Journalists, Educators, Political Aides, and the Interested Public. New York: Columbia University.

Milliken, FJ. . 1987. "Three Types of Perceived Uncertainty about the Environment: State, Effect, and Response Uncertainty. *Academy of Management Review12* (1). 133–143.

Schneider, R. O. .2011. "Climate Change: An Emergency Management Perspective." *Disaster Prevention and Management* (20): 53–62.

European Spatial Planning: Adapting to Climate Events, 2007. "Climate Change Communication Strategy. a West Sussex Case Study." http://www.espace-project,org/part1/publications/reading/WSCCClimateCommunications%20Strategy.pdf.

Population Media Center. https://www.populationmedia.org/about-us/.

Slovic P. 1997. *Trust, Emotion, Sex, Politics and Science: Surveying the Risk-assessment Battlefield.* San Francisco. The New Lexington Press.

Anthony Leiserowitz, 2007/2008. "International Public Opinion, Reception, and Understanding of Global Climate Change." UNDP: Human Development Report.

Wibeck, Victoria. 2014. "Enhancing Learning, Communication and Public Engagement about Climate Change-Some Lessons from Recent Literature." *Environmental Education Research* 20 (3): 387–411.

The EU's Climate Action Campaign. http://ec.europa.eu/climateaction/index_en.htm.
The Swedish Climate Campaign. http://www.naturvardsverket.se/Nerladdningssida/?fileType=pdf&downloadUrl=/Documents/publikationer/620-8153-5.pdf.
Krasner, Stephen. 1983. International Regimes. U. S.: Cornell University Press. The World Bank . 1996. *The World Bank Participation Sourcebook.* Washington. D. C.: Environmentally Sustainable Development.
Keskitalo, Carina. 2004. "A Framework for Multi-Level Stakeholder Studies in Response to Global Change." *Local Environment,* 9 (5): 425–435.
Charles F Parker, Christer Karlsson, and Mattias Hjerpe. 2015. "Climate Change Leaders and Followers. Leadership Recognition and Selection in the UNFCCC Negotiations" *International Relations* 29 (4) .434–454.
White House. 2014. "U. S.-China Joint Announcement on Climate Change." https://www.whitehouse.gov/the-press-office/2014/11/11/us-china-joint-announcement-climate-change.
White House. 2015. "U. S.-China Joint Presidential Statement on Climate Change." https://www.whitehouse.gov/the-press-office/2015/09/25/us-china-joint-presidential-statement-climate-change.
White House. 2016. "U.S.-China Point Presidential Statement on Climate Change [EB/OL]." https://obamawhitehouse.archives.gov/the-press-office/2016/03/31/us-china-joint-presidential-statement-climate-change.
White House. 2017. "An America First Energy Plan." https://www.whitehouse.gov/america-first-energy.
LYNAS M. 2009. "How Do I Know China Wrecked the Copenhagen Deal? I Was in the Room". Guardian, https://www.theguardian.com/environment/2009/dec/22/copenhagen-climate-change-mark-lynas.
Loughran, J. 2016. "China Could Become the New Flag-bearer in the Fight against Global Warming Following the Election of Climate Change Sceptic Donald Trump in the US Presidential Elections." https://eandt.theiet.org/content/articles/2016/11/china-touted-as-next-global-climate-change-leader-after-trump-victory/.
Anthony Leiserowitz, and Edward Maibach, etc. "Climate Change in the American Mind. November 2016, Yale Program on Climate Change Communication." http://climatecommunication.yale.edu/publications/climate-change-in-the-american-mind-november-2016/3/.
Kal Raustiala, David G. and Victor .2004. "Regime Complex for Plant Genetic Resources" *International Organization* 58 (2).
UNFCCC, 2017, "New Surveys Show a Majority of Americans and Chinese Support the Paris Agreement and the Transition to a Low Carbon Future", https://cop23.unfccc.int/sites/default/files/resource/Press%20Release.pdf.
UNFCCC, 2017, "A Majority of the Public in the US and China Support the Paris Agreement and the Transition to Clean Energy, New Survey Findings Show", https://cop23.unfccc.int/sites/default/files/resource/Key%20Findings%20-%20The%20Climate%20Change%20in%20the%20Chinese%20and%20American%20Mind%202017.pdf.
UNFCCC. 2017. "China4C's 2017 National Public Opinion Survey Report Climate Change in the Chinese Mind Released at COP23", https://cop23.unfccc.int/sites/default/files/resource/Press%20Release%20-%202.pdf.
Commission on Global Governance. *Our Global Neighborhood.* Oxford University Press. 1995.
Wang B, Zhou QS. Climate change in the Chinese mind: An overview of public perceptions at macro and micro levels. WIREs Clim Change. 2020; e639. https://doi.org/10.1002/wcc.639.
INDC to Target Partnership. www.ndcpartnership.org.
James N. Rosenau. *Governance without Government.* Cambridge University Press. 1995.
Official website of Climate Bonds Initiative: http://www.climatebonds.net/2016/12/poland-wins-race-issue-first-green-sovereign-bond-new-era-polish-climate-policy.
Oxfam. Report on Extreme Inequality in Carbon Emissions. http://www.oxafm.org.cn/download.php?cid=18&id=208.

Peters Edgar E. *Complexity. Risk and Financial Markets*. Translated by Xuefeng Song and Qingren Cao. China Renmin University Press. 2004.
Tokyo Protocol. http://unfccc.int/resource/docs/convkp/kpchinese.pdf.
United Nations Framework Convention on Climate Change. 1992. http://www.un.org/zh/aboutun/structure/unfccc/0.
United Nations Educational, Scientific and Cultural Organization. Precautionary Principle. 2005.
R. Edward Freeman. *Strategic Management: A Stakeholder Approach*. Translated by Yanhua Wang, Hao Liang. Shanghai Translation Publishing House. 2006.
Website of United Nations Framework Convention on Climate Change. The First Two-Year Update Report of the People's Republic of China on Climate Change. http://unfccc.int/files/national_reports/non-annex_i_parties/biennial_update_reports/submitted_burs/application/pdf/chnburl.pdf. Last visit time: December 2016.
Rossa, and K., R. P. Song. 2017. "China Making Progress on Climate Goals Faster than Expected [EB/OL]." World Resources Institute. http://www.wri.org/blog/2017/03/china-making-progress-climate-goals-faster-expected.

Printed in the United States
By Bookmasters